Graham Rice

# HARDY PERENNIALS

TIMBER PRESS
*Portland, Oregon*

For Sue Jackson

First published in North America in 1995 by
Timber Press, Inc.
The Haseltine Building
133 S.W. Second Avenue, Suite 450
Portland, Oregon 97204, USA

Typeset by Selwood Systems, Midsomer Norton
Printed in Great Britain by Butler & Tanner Ltd, Frome, Somerset
Set in 10/13pt Sabon

ISBN 0-88192-338-9

A few small sections of this book first appeared in *Practical Gardening*,
*Gardens Illustrated* and the *Observer*.

# CONTENTS

# LIST OF ILLUSTRATIONS

# ACKNOWLEDGEMENTS

A huge number of people, knowingly or otherwise, have helped in their various ways in the preparation of this book, some by providing inspiration and support, others by providing information. I would like to thank them all, whether or not they are mentioned in the text.

I would especially like to thank Jean Emmons for her beautiful drawings and John Fielding and Peter Ray for the use of their superb photographs. In particular I must thank Elizabeth Strangman and Graham Gough for their inspiration, their firm opinions and their occasional scepticism as well as for growing and introducing so many good perennials. And thank you, too, to Sue Jackson for again writing such things as 'What on earth do you mean by this?' in the margins of the draft.

GRAHAM RICE
November 1994

# INTRODUCTION:
# EVERY VIRTUE

# INTRODUCTION: EVERY VIRTUE

## The diversity of hardy perennials

'The hardy perennial possesses every virtue that you could require of a plant,' wrote Christopher Lloyd in his book on perennials published nearly thirty years ago. These virtues are indeed so many and so diverse that the whole of this book is needed to enumerate and discuss them.

For there can hardly be a feature or a combination of features which we can reasonably demand and which hardy perennials cannot supply. Even the winter, when so few perennials are in flower, provides a variation in the way in which we appreciate them – for this is the season for individual varieties, even for individual specimens, compared with the woodland tapestry of spring or the bold sweeps of colour in the summer border.

In fact about the only feature which is common to hardy perennials is that all are gathered under that all-embracing heading: every flower colour is available, the flowers may be grouped singly or in hundreds, they may be fleeting or lasting; perennials come in every height from a few inches to over 10ft; most die down in winter yet a few disappear in summer and a few not at all; some have undistinguished foliage while the foliage of others is their only attractive feature; their leaves are generally green, but most other colours can be sought out; some are grown for their stems, fruits, habit of growth or the very structure of the plant itself.

Hardy perennials also come with so many associations: their natural habitats and the ways in which they have come into our gardens; their development in cultivation; the people and the gardens with which perennials are associated; the infinite variety of ways in which they can be grouped together.

This simple outline of the virtues of perennial plants is no more than that; the book adds the detail. But one book cannot be truly comprehensive; here, I have not attempted to be so.

It is only natural to discuss the plants I like best and admire the most, and those which excite my special interest. Being something of a non-conformist, alongside the classic plants like delphiniums and phlox, hostas, hellebores and asters I've particularly included less well-known varieties or groups and those which are growing in popularity or in need of a revival; so epimediums and celandines are featured. I've also looked back to the day when lavish catalogues were bursting with perennials – but not for nostalgic reasons. Rather I seek to emphasize that the extraordinary diversity which once enriched both our gardens and our winter firesides as we browsed the catalogues should enlighten our imagination as we look ahead to new developments and not simply prompt nostalgic reveries.

Plants which are newly bred, newly discovered or which have remained unappreciated in the wild also interest me and I have tried, to some extent at least, to redress a current imbalance by highlighting some American native plants as well as garden varieties of American origin. On both sides of the Atlantic plants from Europe and Asia and even South America have captured our attention and in doing so have tended to eclipse North American natives. We British revel in our horticultural heritage and at the same time

American gardeners often continue to look to Britain as a source both of plants and of inspiration, so attention is deflected from the diversity of the American flora as a source of garden plants.

In Britain, where for hundreds of years special forms of our native plants have been brought into cultivation, new ones are still being found which make good garden plants – and which then find their way across the Atlantic. I realized on my recent visit to the United States that in temperate zones the opportunities are so much greater; the state of Oregon alone has almost twice as many native species as the whole of the British Isles. So by highlighting some American plants, perhaps new discoveries and a reciprocal flow might be encouraged.

Not all the plants are treated in quite the same way; even if I had created a straitjacket of that sort I'm sure I would be breaking out of it in the first chapter. So instead I've chosen to discuss features which seem especially interesting, whether it be the development of varieties, the merits of one variety as compared with another, ways in which the plants are used in the garden or propagation; I have not attempted to be comprehensive or to treat all the entries in a uniform manner. This will doubtless lead to some surprising omissions as well as unexpected inclusions – and to the occasional exasperated reader.

This is a book which features widely popular perennials alongside those which are almost completely unknown, perennials which have been grown for hundreds of years alongside those discovered or bred in the year this book was completed, perennials from both sides of the Atlantic and, incidentally, quite a few plants which are not perennials at all – this simply reflects the way we grow perennial plants in the garden. They do not even have virtue in common, for a few with no such quality seem to have crept in, albeit with appropriate warnings. So this is a heterogeneous book, a book to match the plants.

# Crimes and Misdemeanours

## Names and definitions

Gardeners with a head for detail tend to argue about hardy perennials – or at least about exactly which term should be used to describe them. In theory the term 'hardy perennial' refers to a plant which can be generally relied upon to survive the winter out of doors and which will live for many years unless mistreated. This is perfectly acceptable as far as it goes, indeed it could be said to go too far, for it's clearly a definition which also embraces trees and shrubs.

An alternative phrase is 'herbaceous perennial', but the word 'herbaceous' means dying down to the ground every year and some of the plants we naturally include here are blatantly evergreen. Even within a single genus, the hellebores for example, there are some species which vanish entirely for the winter and some which remain attractively evergreen. And this phrase also leaves out any consideration of hardiness.

These two terms are sometimes combined into one, 'hardy herbaceous perennials', which is unreasonably cumbersome, yet at the opposite extreme phrases like 'border perennials' and even 'border plants' are used, perhaps because they sound less intimidating, although their vagueness allows just about every plant in the garden to be included.

All this produces some entertaining anomalies. Bergenias, with their woody rhizomes, would surely be classified as shrubs if their growth was a little more upright, for they are both woody and evergreen. And vincas, included by most authors as shrubs, have so little woody growth that such classification seems perverse.

And these enigmatic bergenias – if their fat rhizomes grew horizontally like those of bearded irises, some nurseries would doubtless sell them in bags of dry sawdust along with other plants with fat storage organs like irises and cyclamen. They would then be grouped under another term which has become increasingly vaguely applied, 'bulbs'; after all, hostas, polygonatums and hemerocallis have been sold as 'bulbs' in their time.

Whichever of these terms we use, even the most pedantic of gardeners knows what they actually mean in everyday conversation, although the elasticity of these definitions still occasionally produces rumblings from some quarters of the Hardy Plant Society – whose name, paradoxically, covers everything we grow outside in our gardens from larch to larkspur. And expending energy arguing about such niceties seems positively sinful when there are so many wonderful perennials to grow. No, these are hardy plants which do not have woody stems except, perhaps, at ground level, and whose flowering stems die down in winter even if they keep a few leaves until spring.

Hardiness, of course, is entirely subjective. In Britain, where in global terms the climate is fairly uniform, a plant which will thrive in the open in a clay soil at one end of the country may need only better drainage and the protection of a sunny wall to survive winters out of doors at the other.

In the United States not only are the differences in climate far greater, but highly significant variations can occur even within a single state.

However, American gardeners have developed a sophisticated awareness of what 'hardy' means in their individual situations as compared with those of other gardeners, and have also become increasingly astute in their understanding of how hardiness in another state or in Britain relates to their own situations. Generally, though, the plants discussed in this book are hardy in Britain, most of which is in zone 8.

While discussing definitions, I should mention that I have only used the word cultivar when particularly necessary. This may be a mistake, but it is an ugly word and one which is in current use only among the most botanically minded gardeners. The word variety may be less precise and, frankly, incorrect; but I use it nevertheless.

Then there's the problem of plant names. *The Plant Finder*, by listing 60,000 plants together with the nurseries which stock them, is an astonishing achievement in itself but at the same time attempting to standardize botanical names is perhaps asking too much even of plantsman Tony Lord. His researches and those of others of the new generation of garden-wise taxonomists such as Chris Brickell, formerly Director-General of the Royal Horticultural Society, Susyn Andrews of Kew, Alan Leslie of the RHS, Piers Trehane, editor of *Index Hortensis*, and all those who have worked on the *European Garden Flora* have done an enormous amount to bring the nomenclature of garden plants up to date. But this has led to changes to familiar names – and, I regret to say, some prompt changes back.

In this book I have generally followed *The Plant Finder*, except in some of its wilder excesses, and have sometimes added explanations for current thinking and for changes from older names and have re-emphasized long-standing and undisputed changes which gardeners have somehow ignored. Many of the plants under discussion have been grown in gardens for centuries, and the names of some have been changed more than once as research has advanced. Progress towards the ideal of stable names which everyone can accept is inevitably hindered as advancing research reveals unexpected relationships. The problem is exacerbated by differences of opinion not only around the world but even within one botanical institution – even individual botanists have been known to change an opinion, and thank goodness that unlike politicians they can be honest about it. Now I'm pleased to say that plans are being laid to ensure that names change a little less often than in the past, especially when the plants involved are widely grown. But names will always provoke argument.

At the worst, it's true that I have used a few unfamiliar names for familiar plants; but at best I can say that no names have been used which the reader will be unable to track down with ease unless an explanation has been included. Gardeners everywhere are being forced to take more notice of nomenclatural niceties, and many enjoy delving into their intricacies. For names are but labels of convenience, shorthand for a full description of the plants; their usefulness depends on us all being prepared to use the same system. The botanical world is now confessing their misdemeanours of the past and taking a great deal more notice of gardeners; gardeners could reciprocate less grudgingly.

*Opposite*
*Helleborus foetidus* 'Wester Flisk', *Iris lazica*, *Iris foetidissima*, *Anemone nemorosa* 'Robinsoniana', *Galanthus nivalis* 'Scharlockii'.

# WINTER into SPRING

# WINTER INTO SPRING

Winter is not necessarily cold. As I look out in early February the stems of the perennials, left uncut in the hope that they would spend the winter glinting with rime, are soft and dripping with waterdrops. On only eight or ten days since November has the morning glistened with ice. But this included a succession of cold, humid days and harsh, frosty nights just after Christmas which combined to add daily to the lattice of ice on the branches. The result was a spectacular accretion of sharp crystal on crystal until I feared the budded spikes on *Mahonia* 'Lionel Fortescue' would simply snap under the weight.

Now, after a long spell of mild damp days with perhaps just a touch of ground frost at night, all those umbellifers, those grasses and asters look soggy, limp and worthy of nothing but prompt composting. The sparrows, I might say, have snaffled the buds on the mahonia.

This continual exchange of the icy for the insipid is bad for both plants and gardeners. The hellebores can be flat with frost in the morning, although of itself this usually does them little harm. But by afternoon conditions are perfect for the speedy spread of the dreaded leaf spot. Dicentras, delphiniums and other early starters do indeed make a pre-emptive dash for maturity only to be brought up short, or rather frosted off black, as that optimistic succulence is crisped in a night and rots in a day.

Gardeners too are hustled into unseasonable activity. In most gardens mowing the lawn in January is surely a sign of having been too lazy or too preoccupied to mow it in the autumn – it rarely grows much in the short winter days, however mild. That other early favourite, sowing begonias and pelargoniums in January, marks the defeat of reason by tradition. It should soon turn into a victory of experience over blind reliance on half-remembered ways as sowings fail, space is overfilled and plants become pot bound and starved before planting time.

In the last week, peering over fences from footpaths and train windows, I've seen people splitting anthemis, replanting pinks, planting daffodil bulbs forgotten in the shed for months, and even moving crocuses in flower. The days are warm and damp: 'Aha,' they cry. 'What wonderful planting weather!' In truth it is still February and wonderful weather for looking; walk (not on the beds, please), look, gawp in astonishment, cut, photograph and congratulate yourself on your foresight in planting such a variety of winter flowers. Otherwise, 'If in doubt, wait it out' is surely the answer.

### Plant for a winter pleasure

But while restraint is essential in practicalities, we can indulge ourselves in the flowers. For no sooner do the last little yellow daisies of *Dendranthema nankingense* slide away and the quirky pink flowers of the 'Emperor of China' chrysanths succumb to sogginess, than crocuses are up, the bergenias have turned, all those delightful but sometimes unnervingly similar forms of *Iris unguicularis* demand inspection, inhalation and cutting, and what was once known as *Helleborus kochii* is approaching Christmas in flower more reliably than most Christmas roses.

For in Britain at least there are few gardens where it is not possible to grow plants which will produce flowers all winter. Vita Sackville-West is much quoted as suggesting, or perhaps instructing:

> Gardener, if you listen, listen well;
> Plant for a winter pleasure, when the months
> Dishearten...

Unfortunately she then spoils that fine and eminently sensible advice by introducing the 'brittle violin of frost'. But she is certainly right. Sissinghurst is not open in January, so we cannot check for ourselves whether or not she followed her own advice. Those of us who write about plants, myself included, propagate far more ideas than we can ever put into practice; this does not mean, of course, that they are necessarily bad ideas.

But how many gardens are both featureless and colourless in the winter? And counting forsythia is cheating – in Britain winter is December, January and February; March is definitely spring. After the Michaelmas daisies have faded, many gardeners blindly abandon the garden until they notice the snowdrops as they hurriedly scrape the ice off the windscreen when late for work. But not, I trust, readers of this book.

### Winter warmers

For the more thoughtful gardener, pleasure in the winter garden is especially poignant and encouraging, although the shrubs and perennials, alpines and bulbs which can open our wan hearts to winter without admitting the chill are relatively few. But there are certainly enough to spark a flash of excitement. So we should cherish them and provide them with the conditions they appreciate.

Good shelter is perhaps the most important factor in ensuring their contentment: shelter from the bitter winds, even shelter from breezes which will waft away that unseen haze of warmth and fragrance. And shelter too from early morning sun, for a rapid thaw of frosted flowers can reduce them to pulp when a more tender and coaxing warmth would allow them to adjust gently.

Many of these earliest and toughest of perennials need shrubby companions for a change of height as well as shelter and a variation in style. Witch hazels are essential for flowers and for the sharpness of their scent. The red-flowered *Hamamelis* x *intermedia* 'Diane' with the white stems of *Rubus biflorus* fronted by the stems of dogwoods and willows in red, green, yellow and dusky purple makes a scene sparkling enough in itself even before the addition of the red stems, dissected foliage and green bells of *Helleborus foetidus* 'Wester Flisk'. Add, perhaps, the bold red foliage of *Bergenia* 'Eric Smith', carpets of the classiest snowdrops you can run to, or even the humble 'Atkinsii', plus winter aconites and palest blue *Scilla mischtschenkoana*. A January and February scheme to match anything from June.

# Hellebores

Choosing the hellebore as the first individual plant to consider in this book is perhaps fortunate. For in many ways hellebores epitomize the delights and the desirabilities, the mysteries and the muddles to be met with in discussing all the plants that follow.

It is true that there are plants whose character is enhanced by a little mystery or curiosity in their background. Indeed some, boasting but little in the way of ornamental features, rely on such curiosities to engage our interest: mistaken histories, confused nomenclature, arguments over distinguishing features hardly visible to even the discerning gardener and misunderstandings between botanical correspondents long dead. Naturally, these are largely excluded from a book which is, or at least purports to be, about plants which are worth growing because they are beautiful.

### Delights and desirabilities

It is true to say, as many a writer has done before me of course, that hellebores would be no less captivating were they to flower in June and July. But the excitement of their cautious emergence not so very long after the rest of the garden has been frosted to collapse, when our senses are roaming restless for a place to pause, does imbue them with a special quality; hellebores in December, the first flowers of spring when autumn is barely over.

'Praecox' (meaning very early) was a form of *H. niger* once grown in Europe, and perhaps this was the one referred to by Canon Ellacombe as being 'now so popular that we have in different parts Christmas-rose farms, from which the fine flowers of the purest white (otherwise unsaleable) are picked by tens of thousands in the fortnights before and after Christmas'. But although 'Praecox' is listed still in Germany, in recent years there have been only sporadic reports of any Christmas roses actually flowering out of doors at Christmas. Some sightings, on investigation, refer to plants which have been recently moved and so jolted into flower. Others produce timely flowers one year but not the next; only a very few seem reliable.

In recent years I've been growing one called 'Higham's Variety', introduced by Carol Klein, but it seems so weak that although heartening in its inclination to flower early, it takes so long from the first hint of the buds at ground level to their final opening that other plants, not to mention the slugs, overtake it. One which sounds all too reliable is 'Trotter's Form', raised in Inverness by Dick Trotter. This was said to flower for nine months of the year, an interesting prospect and guaranteed to amuse visitors I suppose; but quite unnecessary.

For earliness, for Christmas perhaps or at least for that tantalizing glimpse of the future, three other plants are certainly more dependable than any form of *H. niger*, and these are the two which gardeners have long called 'Atrorubens' and *H. kochii*, plus the Corfu form of *H. odorus*. At once the mysteries and muddles are upon us, for neither of the first two names has any right to be used for the plant to which it is attached. The plant we all know as 'Atrorubens' has nothing to do with true *H. atrorubens*; *H. kochii* is not a valid species; what's more, that *H. odorus* may actually be *H. cyclophyllus*.

To help resolve the first confusion Brian Mathew proposed the name 'Early Purple' for 'Atrorubens' and its placing under *H. orientalis* subsp. *abchasicus*; this is sensible, but there seem to be at least four different forms of the plant so now, following the recent tendency to use Group names and with no advice from taxonomists to the bewildered, not to say irritated, gardener as to the definition of this concept, they are classified as Early Purple Group. These are all reliably early-flowering forms with flowers in various degrees of purple, some with a noticeable amount of green.

The best of these are invaluable plants and good seedlings can be raised, like The Plantsmen's 'Aldebaran'. Although sometimes said to be sterile, the unhelpful weather and the absence of pollinating insects at flowering time are the

Unfortunately, these perfectly evenly spotted forms of *Helleborus orientalis* rarely come true from seed

significant factors, rather than any genetic sterility. Grow them uncrowded by other flowers except perhaps for a background of evergreen shrubs and a carpet of early snowdrops tucked around their feet.

*H. kochii* is an invalid name for a pale, early-flowering form of *H. orientalis* and has sometimes been treated as an Orientalis Hybrid and known as 'Kochii'. Usually creamy or almost white in colour, often flowering in December although seedlings differ, the variability of this plant has again induced taxonomists to resort to the Group concept; Kochii Group is now the name. An evergreen daphne behind and red bergenia foliage alongside could complete the picture.

The Corfu form of *H. odorus* has been distributed by Blackthorn Nursery and is a truly dependable early plant with luminous green flowers over many weeks. The slightly greeny silver foliage of the winter *Elymus magellanicus*, the milky chocolate *Carex comans* 'Bronze Form', with snowdrops or 'Cedric Morris' daffodils are good companions. There is such doubt about the distinctions between *H. odorus* and *H. cyclophyllus* that assigning any single garden plant to one species or to the other is a risky business. I will flow with the tide and call it *H. odorus*.

Hellebores have long been loved for their winter flowers; more recently the ever-increasing range of pure colours and elegant spotted forms developed by Elizabeth Strangman, Helen Ballard, Robin White, Will McLewin and Kemal Mehdi have captured our attention. And the wild species, demure and refined by nature, have met a developing taste for plants in a more restrained style. At ease with these advances I was content to lie back relaxing in the steady flow of companionable change, but now I struggle against the current; complacency could never last, I suppose.

For double hellebores have surged into favour, indeed they are proving so singularly desirable among gardeners that the undignified scrum and scuffle to buy them at the rare shows at which they appear rather debases their appeal; they have even been dug and stolen from gardens. As it happens, I cannot say that I take much delight in most of them. Those misty purple and green doubles, 'Dido' and 'Aeneas', the forms of *H. torquatus* Elizabeth Strangman found in Montenegro in 1971, are lovely; modest, demure with a subtlety of colouring that the more recent crosses between them and the coloured Orientalis Hybrids conspicuously lack.

A few of these new crosses *are* highly desirable; an enchanting double smoky purple with the petals laid evenly one over the other like fish scales is one, and another in a similar style is 'Granny Smith' apple green. But those spiky, spidery ones like fluorescent crabs are more than I can take.

These doubles seem to have taken over top rating on the drool scale from spotted forms, or at least those spotted forms where the spotting is distinct and even, rather than muddled and messy. I have a pale primrose seedling with a neat, sharp speckling of red covering about half of each petal which is simply exquisite. And another greenish yellow seedling with purple nectaries and a neat, triangular purple mark at the base of each petal which left me gawping when I turned up its very first flower. And there's a seedling raised by Kemal Mehdi, illustrated in my hellebore book, a pure white, completely and evenly spotted in red. Words fail me.

And this is a delight as penetrating as the drawing forward of spring into winter by the earliest forms. Turning up a flower to reveal its unique colouring may not always prompt a happy surprise but when it does, innocent visitors must be led through rain and hail and forced to their knees to inspect the latest wonder.

## Mysteries and muddles

How much time have you to spare? I could fill the book with them. (Come to think of it ...) But rather than discuss the minutiae of intro-gression and infraspecific variation, let me outline for you the general nature of the problems which afflict hellebores the better to understand specific difficulties when they arise.

Most of the difficulties, both in the wild and in gardens, arise in the group which includes *H. orientalis*, *H. odorus* and *H. cyclophyllus*, *H. multifidus* and *H. torquatus*, *H. viridis*, *H. atrorubens* and *H. purpurascens*. One problem creating unnecessary confusion is the use of old or invalid names. As just discussed, *H. atrorubens* is a name which must only be applied to the wild species from a small area in Croatia and Slovenia and not to the garden hybrids previously referred to by this name. The continued use of names like *H. colchicus* and *H. olympicus* only serves to encourage an impression that these doubtfully distinct forms are true species, while at the other extreme using the name *H. orientalis* to describe all the garden hybrids only fuels a mis-understanding as to the true hybrid nature of the garden plants. For in gardens this whole group is largely represented by what are increasingly known as the Orientalis Hybrids, plants deriving from crosses between *H. orientalis* and one or more of the other species. They are also known as *H. x hybridus* but this name, although a valid one, is far too wide and imprecise to be altogether helpful.

There is still a little confusion about the well-known and invaluable *H. argutifolius*, listed in catalogues even now under the invalid name of *H. corsicus*. Fortunately, the names of its best-known hybrids, *H. x sternii* (*H. argutifolius* x

*H. lividus*) and *H. x nigercors* (*H. niger* x *H. argutifolius*) are well set in our minds; now we must only get to grips with the last hybrid with *argutifolius* blood, *H. x ericsmithii* (*H. niger* x *H. x sternii*).

Apart from problems with names, fecundity and variability cause most of the confusion. First, individual species are themselves variable and features such as flower size and flower colour, leaf shape, colour and hairiness of new foliage, and the number and shape of leaf divisions may vary from one plant to another – or sometimes a whole population will be noticeably uniform. In Britain *H. viridis* is much the same wherever you find it, in Serbia *H. torquatus* is rather the opposite – populations are distinct from each other and within a single population plants are noticeably variable, hardly two may be similar.

Then there is the familiar misconception that species plants always come true from seed, whereas in truth they are not only largely inter-fertile, but actively promiscuous; the stigma of each flower ripens before its pollen specifically to encourage cross-pollination. So growing any two species in the same garden can produce hybrids aplenty, and these hybrids may be inter-mediate or one parent may dominate and the offspring show relatively few characteristics of the other. This intercrossing also occurs in the wild where different species grow within bee-reach of each other. In some populations of *H. odorus* plants with bronzed green flowers are found, the result of hybridization with *H. tor-quatus* growing near by. So the belief that the species come true from seed soon causes chaos, leading to a wide variety of plants which are not at all similar to their seed parents being circulated under misleading names.

If anything, the Orientalis Hybrids are even less reliable in this respect. Unfortunately, even when bees or the gardener self-pollinate a single plant, the genetic make-up of many is so muddled that even these offspring are different from their parents. About the only thing you can depend upon is that if you have got spotted plants in the garden and you allow bees to pollinate and the self-sown seedlings to flower, sooner or later all your plants will have spots.

In practice, there is one particular problem which results from this combination of fecundity and variability. You buy a named plant, one of Helen Ballard's or Elizabeth Strangman's perhaps, 'Philip Ballard' or 'Dido'. You collect seed from it and raise the seedlings, then you give a few away to friends; but the name of the plant from which seed was collected tends to go with its seedling, though it is certain that the seedling will not be identical to the parent. Hence the many plants claiming to be named forms which are not true. The logical next step is to consider propagation.

### Division and seed

The only way to be sure of replicating a plant exactly is to divide it; seedlings will not come true. Division of the Orientalis Hybrids is best done in late summer and early autumn, shortly before the roots naturally start a burst of growth. If the intention is simply to expand a single plant to a group of three, the traditional two-forks technique for dividing perennials will work well enough. But hellebores always take a long time to settle down after division; their root system is so deep that the majority is left behind however carefully the plant is lifted. So perhaps it is no bad thing, while upsetting the plant so much, to divide it up into a larger number of smaller pieces which will give you more plants in the long run.

Lift the plant, wash off all the soil using a hosepipe, and with a stout kitchen knife (I use an old bread knife) and a pair of secateurs cut the plant up into pieces with just one or two noses on each. Trim off any old woody parts plus the longest of the old, black roots but retain as many of the paler and especially the new white roots as possible. Try to ensure that each piece retains at least one healthy leaf.

Larger pieces can go straight back in the garden but all are best potted into 5in pots and kept in a shady frame until they become established. When the roots peep through the bottom of the pot they are ready for planting.

There is one great myth about raising hellebores from seed, and that is that the seed should be kept until late autumn or winter and sown then to benefit from winter frost. No. Seed is best sown fresh in June or July and will usually come up like cress during the winter if left outside in a frame and never allowed to dry out. Space sow the seed; about twenty-two seedlings will go in a 5in pot. Prick out the seedlings into $3\frac{1}{2}$in pots just as the first true leaves are showing in spring, then pot them on or plant them out when roots appear through the holes in the base of the pot. This is a perfectly reliable and simple method, and again I emphasize the importance of sowing in summer so that the seed receives a warm, moist period before the temperature drops.

Seedlings, as I have reiterated more than once, will vary. They will usually vary less if the flowers of a treasured plant are self-pollinated by hand and the best of the resultant seedlings kept for planting in the garden – always bearing in mind that these do not deserve the same clonal name as their parent.

### Verse and verse

I would not at all wish you to gain the entirely false impression that every chapter will end as this one does, but I could not resist.

'To be fragrant,' wrote Colette, 'is not its mission, but let December come, let wintry frost blanket us, and the hellebore will show its true colours. A nice deep snow, not too powdery, a little heavy, and winter nights that the west wind passes through like a precursor, now that's what makes a hellebore happy.' Hmmm.

What a relief, then, that we have the poet whom Canon Ellacombe describes as 'scarcely worth quoting': C. Mackay, what an excellent fellow, writing in *Punch* in December 1882. Are you ready?

> *Know ye the flower that just now blows,*
> *In the middle of winter – the Christmas*
> *rose –*

Not exactly Wordsworth or Housman or John

Clare, is it? He goes on . . .

> *Though it lack perfume to regale the*
> *nose,*
> *To the eyes right fair is the Christmas*
> *rose –*
> *A fiddlestick's end for the frost and the*
> *snows;*
> *Sing hey, sing ho, for the Christmas rose.*

From now on I expect the whole country to be roused from its favourite soap opera every winter by delighted cries of 'Hey' and 'Ho' when the first flowers of the Christmas rose appear. Even if this be not until February or March.

# Irises

In May and June, it's perfectly possible to be overwhelmed by irises. Those glorious colours and that heavy fragrance swirl over you, numbing the senses so that the individuality of the many, many varieties may never quite sink in. They provoke a broad, sweeping response and it is easy, taking the trouble to look at an individual, to be distracted not only by immediate iris neighbours but by the idea of the whole clan. A particular response always seems hazy and faint at the edges. Not that discrimination is impossible, it just needs working at. It demands an almost structured technique of appreciation. You single out the bearded irises from the Siberians and the Spurias; then the tall types from the intermediates and short; the blowsy from the neat and so on. But winter is different.

### If you could choose just one plant...?

Winter is simple, for there are just two very distinct border irises and what is generally known as the winter iris creates a very individual feeling of excitement and protectiveness. Forming ever-increasing clumps in gritty soil at the foot of sunny walls, the dark narrow leaves often become ragged and it pays to follow Sir Frederick Stern and cut them back by a third in late summer. The slender pointed buds are like Technicolor umbrellas, and when they unfurl the Algerian or winter iris, *I. unguicularis*, is a showstopper indeed. Selected by E. A. Bowles as the first flower of spring, from October onwards the flowers appear, purplish and true blue, lavender or white, building up to a crescendo in February and March when you can gather enough flowers from a single clump to enjoy indoors and leave the plant in the garden apparently unplundered.

There is an art to cutting these flowers. In this iris, what appears to be the stem of the flower is actually an unexpectedly elongated part of the flower itself and only at ground level is there $\frac{3}{4}$in of true stem. If you simply slice off the flower as low down as possible with a sharp knife or secateurs you will have committed a great and destructive folly. For that stub of stem usually carries not one but three flowers, the one you are keen to pick and two more still in tight bud, right down at the base. Cut as low as possible and those two embryonic buds will also be cut away and will never bloom. So to avoid the carnage slide your finger down the stem, part the sheaths at the base until you reach the other waiting buds, and cut just above them. Do this just before the buds open and bring them into the house to watch the flowers unfurl and feel the warm scent tickle your nose.

There are four varieties commonly grown. One is usually available simply under the species name and is lavender in colour. 'Mary Barnard' is dark blue, 'Walter Butt' is a soothing, cool, pale, pearly lavender and reliably early, and there is 'Alba', the white form, which varies rather but tends to be shy flowering, perhaps owing to virus infection. Seedlings from 'Walter Butt' occasionally come white. But there are more: *Iris unguicularis* var. *angustifolia* has shorter, narrower leaves, *I. lazica* has broader leaves, while over the years many forms have been named, including 'Miss Ellis', 'Peacock', 'Bridal Pink', 'Winter Snowflake' and more. If any of these are spotted on your travels around the country's nurseries, snap them up, for they might just give you a new shade or a different period in full flower.

If you are fortunate enough to have a suitably warm border, perhaps just a narrow 2ft strip on the south side of the greenhouse where a little warmth will keep roots cosy in winter, you have the opportunity to create quite a feature. *Iris unguicularis*, of course, is indispensable and in autumn will overlap with the last of the *Nerine bowdenii*, preferably the variety 'Mark Fenwick' which has more flowers in each head and is more vigorous. The early flowers may also blend with those of *Amaryllis belladonna* (the hardy version, not those indoor monsters) and the white, crocus-like *Zephyranthes candida*, and there must be agapanthus for summer. I would be inclined to add a small annual that would self-sow in such a choice spot, the creamy *Platystemon californicus*, to wander among these other tender treasures and peep its prettily butter-spotted flowers through the gaps.

Once friends see this they will want pieces of the irises without doubt. August is moving time for *Iris unguicularis*, but splits usually take a year or two to settle down, sometimes growing well for a while but hardly flowering. This species can be raised from seed and the pods, bursting with seeds, will be found attached to those short stems down at ground level; the seed can be sown in spring in pots of suitably gritty compost.

'Suppose a wicked uncle,' inquires Mr Bowles, 'who wished to check your gardening zeal left you pots of money on condition that you grew only one species of plants: what would you choose?' 'I should settle on *Iris unguicularis*,' replies Bowles, and considering the extraordinary variety of plants he grew at Myddelton and saw in the wild, we should respect his choice. And if we are without a south-facing wall in front of which to plant our treasures, we must build one forthwith.

### The snake fiddles

Our Mr Bowles is less fulsome, but still enthusiastic, on the subject of the other iris for winter, *Iris foetidissima*: 'few can rival the beauty of the newly cracked pods,' he suggests, without over-extravagance. Singularly slandered by the name stinking iris, the leaves smell but slightly and then only when crushed. The alternative name gladdon may sound weird but in theory is more suitable, deriving from the Latin and Old English meaning 'little sword'; it certainly beats 'snake fiddles' and 'blue seggin' which have also been heard – though quite possibly only by rustic extras in films of Thomas Hardy novels.

This species is widespread in southern Europe and even stretches into North Africa, where especially luxuriant forms are said to grow. I should like to see them; in Britain it is confined to the southern part of the country. The flowers have a quiet beauty – which is code for their being so small and dull that you have get down on your hands and knees to see them at all. This is perhaps true of the flowers of the usual wild form, which are a rather murky ochre in colour with purple tinges, but there are yellow-flowered forms like var. *citrina* from Dorset, in a clear ochre with brown streaks, and even a startling pure yellow, var. *lutescens*, from Algeria. And rumour has recently reached me of a form with pure white flowers.

But interesting as the flowers are, it's the fruits for which we grow this species; quite a change for an iris. In most wild forms the seeds are orange-scarlet and as they burst out of their pods are a truly sparkling winter attraction. There is a form with orange-yellow berries, one with straw-coloured fruits and a spectacular one with pure white berries found in Somerset in 1921. Unfortunately this latter form, known as 'Fructo Albo', produces fewer flower stems than more common forms but I suspect this may be due to virus infection, especially as there are yellow streaks in the deep green foliage. I'm raising some seedlings in the hope of finding a floriferous and virus-free clone but the word from the iris experts is that the variegated form of this species provides the reservoir of virus infection, which is partly why it flowers and fruits so poorly. The virus does not cause the dramatic variegation, but weakens the plant then infects other irises; they all succumb in the end. So perhaps my fat clump of 'Variegata' will have to go.

I must say that having grown 'Variegata' for quite a few years now I like it less and less; at almost every season it seems about to give its best but then rarely manages to fulfil the promise. It looks best so completely surrounded by other flowers that it merges into the broader picture. Never grow it as a specimen.

There are two great things about all these gladdons, apart from their interesting flowers and bursting pods of glittering fruits. First of all they will thrive in less than perfect conditions; in the wild they often grow on dry, chalky soils (in the shade, too, more often than not) and they are as tolerant in gardens. Even 'Fructo Albo' seems to take to relatively inhospitable spots. But just as important is that the foliage makes such dense clumps, and in such a rich, deep colour; rather like English racing green from the days of Jack Brabham.

### The discriminating iris grower

This discussion of just two species from the great family of irises may suit this winter chapter, but it cannot do the genus justice. This is something of a mixed-up book, in the sense that it addresses all the plants of one genus in the one place even if this drags some of them out of season, rather than treating the same genus in two or three different places according to the flowering time of the various types.

It could be argued, I suppose, that the irises of June are more important than the irises of winter and early spring, but are they more valuable? Had as much concentrated imagination and dedication been devoted to those winter irises as has been lavished upon the bearded irises, is it not likely that we would be enjoying a far greater variety of forms?

But for sheer opulence the bearded irises cannot be surpassed, and it's hardly surprising that Monet (and of course Cedric Morris) should have enjoyed and grown so many varieties. But irises have changed since Giverny was made. Now American hybridizers in particular are striving to create new varieties – I hesitate to use the word 'improved' because I do not see all these developments as changes for the better. As I said at the start, discrimination is vital, and not only when the gardener chooses varieties, but when the breeder chooses features to develop.

Selecting for increased branching to give more flowers on each stem seems a sensible approach to take if combined with sufficient increased stability to ensure that plants do not fall over in gales. Using an increase in the number of flowers to ensure a longer season is more valuable than aiming for a short but more flamboyant one. Remontant habit, a repetition of blooming in August following the main display in June, I can do without; fortunately there are but few claiming this feature.

Caution is necessary when breeding to increase the size of the flowers (some can be 7in across and 4–5in high), for large flowers also add to instability especially in wet conditions. When flowers are larger, the petals themselves need to be thicker in texture to ensure that they hold their shape rather than collapse under their own weight.

One trend of which I particularly disapprove and which arises from the demands of the show-bench as opposed to the requirements of irises as garden plants is the tendency to select varieties whose falls – the three flouncy bits at the sides – are held out horizontally rather than hanging down in the traditional manner. This elevation of the falls may almost create a circular platform from which the three narrow, upright standards can erupt, but in the garden the absence of the vertical plane of colour reduces the impact of the flowers unless they are viewed from close to. Monet would never have painted them if their falls had ceased to fall.

The bearded iris is but one of a number of plants where the conventions of the showbench have inhibited breeding developments which would have been of great value to the gardener. The laced polyanthus is another example; for so many years chocolate rimmed with gold was the only colour combination seen, when a variety of other combinations would be well worth having. Some breeders are now working on them.

One trend in iris breeding of benefit to gardeners has been the development of bearded irises of shorter stature than the old tall varieties. The smallest of these, at less than 8in, also flower in May, before the traditional bearded irises, and so extend the season enormously.

Unfortunately for gardeners, the American Iris Society has invented a categorization for bearded irises, partly based on height, which causes as much confusion as clarification. Some catalogues simply divide these irises into dwarf and tall and give the height, which is about all we gardeners need. The exhibitors once had a simple tripartite division of the bearded irises: dwarf, intermediate and tall. Now we have six divisions: Miniature Dwarf Bearded (MDB), Standard Dwarf Bearded (SDB), Intermediate Bearded (IB), Miniature Tall Bearded (MTB), Border Bearded (BB) and Tall Bearded (TB).

This sounds simple enough – you might think it a clear and sensible classification representing a gradual increase in plant size from the MDBs at a maximum of 8in in height to the TBs at 27in or more, together with a corresponding progression in flowering times; be not fooled by the apparent logic of such an arrangement. For the stipulated heights of the IB, MTB and BB irises are the same, 16–27in; there are other differences.

The IBs (in the iris world you must get used to these acronyms) have flowers 4–5in across and flower between the SDBs and the TBs. The MTBs have flowers whose combined height and width (real showbench stuff, this) is not more than 6in and flower with the TBs. The BBs also flower with the TBs but have flowers 4–5in across compared with the TBs at 4–7in. So if I bred a new iris which was 3ft tall and uniquely well branched but whose flowers were only 3in across, and this might well be a superb garden plant, it would not seem to fit into the classification at all.

Of course this is all something of an oversimplification; stalwart exhibitors belonging to the British Iris Society debate the system with a great deal more precision. But it does seem that enthusiasts for *Iris sibirica* and its hybrids have adopted an approach which has a little more relevance to gardeners. It's true that breeders of Siberians have tried to create varieties whose falls stand out horizontally, and indeed Amos Perry succeeded as long ago as the 1930s with 'Perry's Blue'. Now some also aim for standards which are splayed and flattened, but in response Jennifer Hewitt, who breeds Siberians in Shropshire, has coined the phrase 'garden visibility' to remind breeders that horizontal falls and standards do not show up when plants are viewed from the side, as they usually are in gardens.

Most bearded irises are tetraploids, the plants having twice the normal number of chromosomes, but Siberians come as both diploids, the natural form, and tetraploids. The first tetraploid Siberians were created by the American hybridizer Currier McEwen using seeds treated with colchicine, which induces doubling of chromosomes. Tetraploid Siberians tend to have thicker leaves and larger, more substantial flowers with thicker petals.

Two breeders, one American and one British, have outlined their aims in creating new Siberians be they diploid or tetraploid. Among Currier McEwen's aims are new colours, interesting colour patterns, improved branching and bud count, earlier and later flowers, continuing and repeat flowering, very short growth, fragrance and disease resistance. Something of a utopian, is Mr McEwen. British breeder Ray Jeffs looks for grace, flowers held high above the foliage to

improve visibility, strong straight stems, and he likes new varieties to retain the fundamental characteristics of *Iris sibirica* and not reflect developments in bearded irises. He also likes them to be easy to grow, reliable flowerers and without any specific cultural quirks. Good man.

This awareness by iris specialists of features which are of such value to gardeners is welcome news for those of us who simply want irises which are good garden plants for our mixed and herbaceous borders.

So now, perhaps, comes the time to suggest a few varieties. In the bearded irises in particular there are an extraordinarily large number to choose from and I grow what I realize is rather

Raised in the United States about twenty years ago, the prettily patterned *Iris sibirica* 'Flight of Butterflies' is now widely grown in Britain

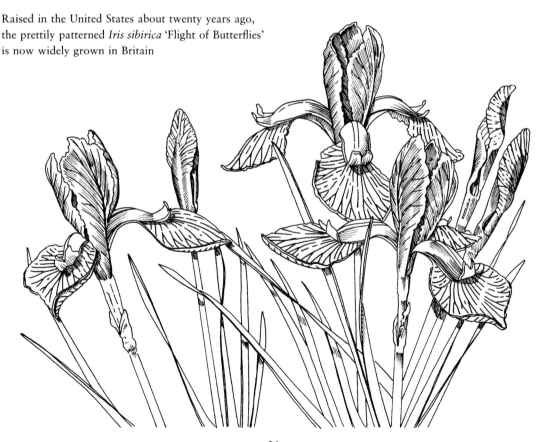

21

an eccentric collection. First I will pick out two old varieties. The tall bearded iris 'Jane Phillips' was introduced in 1950 and is still very much with us; indeed it seems to have been grown less in the 1950s than it is today. It's the most delicate of blues with a white beard and is so floriferous, so tough and so amenable that I could not be without it. However, I must digress here and mention 'Cannington Bluebird', Cy Bartlett's award-winning successor in a lovely soft blue.

Among the Siberians, 'Flight of Butterflies' looks as if it might be a very old variety but it was raised by Seattle iris breeder Jean Witt and introduced only about twenty years ago. Its habit is more open than many, with the flowers dancing at different heights rather than held in one plane. The buds are exquisitely veined in black and open to small, veined blue flowers; the name is most apt. 'Flight of Butterflies' is also a great deal shorter than many Siberians yet it has retained its elegance.

Of the shorter bearded irises (I will be no more specific than that) I am especially fond of 'Cherry Garden', with deep maroon flowers enlivened with a purple beard, 'Green Spot' in ivory white with a green mark on its falls, which are horizontal and actually improve the garden visibility as the plant is only 9in high, and 'Westwell' in violet and white, also with horizontal falls.

In recent years I've made a point of examining the trials at Wisley each year to assess the very latest introductions, some not even named, alongside well-established varieties. The following have especially stood out among the tall bearded irises: 'Blue Eyed Brunette', which in spite of its falls not really falling all that much is a lovely colour, a pale chestnut with a blue flash in front of the beard; the startling 'Dusky Dancer', with black buds, almost black falls and deep purple standards; 'Jitter Bug', with orange-yellow standards and yellow falls heavily spotted in **neat** rows with rusty maroon.

There are fewer Siberians, both in the trial and generally, and also in the range of colours, so I can pick a favourite from each colour range. Pale blue: 'Ian', with large flowers in a vibrant sparkling blue. Mid blue: 'Silver Edge', with silver lacing to its broad, admittedly horizontal falls. Dark blue: 'Germanhet' in deep royal blue. Purple: 'Ruffled Velvet' has flared standards but its purple falls with yellow veins are lovely. Red-purple: 'Ewen', with broad falls. White: John Metcalf's 'Forncett Moon', in a slightly off-white with a yellow throat.

So much included yet so much left out. The Spurias are unjustly neglected both in gardens and here, *Iris chrysographes* in its various forms is a truly beautiful plant, and for those on acid soil there are the neat Pacific Coast irises. Japanese irises, the white variegated *Iris ensata* 'Ghost', and the forms of *Iris pseudacorus* should have been mentioned, as should such oddities as 'Gerald Darby' with the red tint to the base of its foliage and the wonderfully veined yet enigmatic 'Holden Clough'. But there is no room.

---

*Opposite*
This dark blue, almost violet form of *Iris unguicularis* was collected in Algeria by Mary Barnard of Honiton in Devon and is named after her. It was introduced by E. B. Anderson, who said it was the most prolific form.

*Overleaf (left)*
ABOVE Of the hundreds of irises on trial at the RHS garden at Wisley, the recently introduced tall bearded iris 'Jitter Bug' was especially striking for its intricate markings.

BELOW *Iris foetidissima* is unusual among irises in being grown for its berries. They are usually orange-scarlet in colour, but white-berried forms are also found along with this pale form from Four Seasons Nursery in Norfolk.

# Hepaticas

The genus *Anemone*, as created by Linnaeus in 1753, was something of a broad umbrella, sheltering not only plants as diverse as *Anemone nemorosa* and *A. japonica*, which even now somehow retain kinship under *Anemone*, but also *Hepatica* and *Pulsatilla*. This catholic assembly has long since been broken up, but I have a creeping necrotic realization that plants perhaps better considered as alpines are taking over this spring part of the book so I hereby make a summary and not entirely logical decision to discuss hepaticas and to leave out pulsatillas altogether.

---

*Previous page (right)*
ABOVE 'Collarette' is a celandine which makes a leafy mound of prettily marked foliage studded with tightly double flowers, which develop this green eye as they age.

BELOW The intensity of the red winter leaves of *Bergenia* 'Eric Smith' varies a little from year to year in the author's garden, but whether carmine-red or crimson they look well in the morning frost.

*Opposite*
ABOVE Seedlings of *Hepatica nobilis* can be very unpredictable, but this prolific plant growing at Washfield Nursery has large flowers in a good colour.

BELOW LEFT *Helleborus × nigercors* is an invaluable hybrid between *H. niger* and *H. argutifolius* with a dwarf habit, vigorous growth and an unusually long flowering season.

BELOW RIGHT The tough and easy to grow *Cyclamen coum* comes in many forms. Here, at the RHS garden at Wisley, the combination of silver-patterned leaves and dark-stemmed, rich pink flowers is especially effective.

## The many or the few

There are said to be ten species of hepatica – I was surprised too. But it turns out that this mystifying figure quoted in the new *RHS Dictionary* includes species from Korea and China not yet in cultivation, including the yellow, yes yellow, *H. henryi*. So for clarification of the subject we must turn, with confidence, to Michael Myers, who in his articles on hepaticas in the *AGS Bulletin* and *The Garden* discusses the two clearly defined species and one highly variable one.

To summarize, *Hepatica acutiloba* grows in eastern North America from Quebec to Missouri and usually in moist, limestone woods. It has noticeably sharply lobed leaves, with pointed lobes, and Don Jacobs of Eco Gardens in Georgia has made some good selections of this, including a double white called 'Eco White Fluff'. *Hepatica transsilvanica* is a tetraploid which grows only in Romanian conifer woods, has a distinctive running habit which makes it easier to propagate but leads to a less concentrated display of flowers and is larger in all its parts than the most familiar species, *H. nobilis*.

Michael Myers conveniently divides this third, most widespread species into three. In Europe there is var. *nobilis*, a plant which grows in woodlands all over Europe except Britain and the hot and cold climatic extremes. Even this European form is variable, in leaf shape and leaf patterning, not to mention flower size and flower colour. In the United States there is var. *americana* – *H. americana* in the new Alpine Garden Society encyclopaedia and *H. armenica* in the

*RHS Dictionary*, where the otherwise useful page of illustrations is ruined by sloppy captioning and misspelling. This variety grows from Nova Scotia to Minnesota, in drier, acid woods, and Don Jacobs has made some selections of this too, including the deep blue 'Eco Indigo'. In Japan there is var. *japonica*, which seems an exceptionally variable plant with many synonyms and many cultivated forms in Japan. In addition *H. maxima* from Korea, with large hairy leaves but undistinguished flowers, has recently been introduced.

There is also one well-known hybrid, *H.* x *media*, a cross between *H. nobilis* var. *nobilis* and *H. transsilvanica* which occurs in the wild when the two parents grow together and in gardens. It is sterile.

### In dappled shade
Hepaticas are woodland plants; in the garden they appreciate dappled shade, a leafy soil and good drainage, which is not difficult to provide, partly because it is not required on a grand scale. A deciduous shrub which is not too dense can provide the shade, or your hepaticas can go on the north side of an evergreen shrub like a daphne. They will even grow in full sun if they never dry out. They're tough, frost will not harm them. The tall, rather thin shade of old oak trees is perhaps an ideal, or in a new garden a shade house made of timber laths can provide perfect shade.

Some species prefer limy soil while some prefer acid conditions, but I'm sure you will already be aware of your soil's pH and have chosen your species accordingly. Mind you, H. Lincoln Foster, who was for sixteen years President of the American Rock Garden Society, believed that neither in the wild, in his part of Connecticut at least, nor in gardens was pH worth the slightest consideration. Leaf mould plus, perhaps, grit are often useful additions in the creation of hepatica-friendly conditions; it is easy to mock the ideal of a well-drained, humus-rich soil but it is not difficult to create if preparation is thorough. Titivating the soil with a handfork and adding a sprinkling of peat is hardly sufficient.

Planting depth seems important, and the crown of the plant should be allowed to work itself out of the soil slightly. If replanted more deeply flowering becomes sparse, although eventually *H. nobilis* will work its way up to the surface again.

Hepaticas are best left undisturbed for some years after planting. *H. nobilis* and *H. acutiloba* tend to make tight compact plants densely covered in flowers, while *H. transsilvanica* with its creeping habit develops into a more open plant. The hybrid *H.* x *media* is, usually, somewhere between. Eventually flowering of some plants of *H. nobilis* may become rather sparse and at this stage action is called for. F. C. Puddle, the first of the family to be head gardener at the garden at Bodnant in North Wales, had rather a drastic prescription for this malady. He took a spade, sliced vertically through the centre of the clump, then forced a slate down the crack. This was said to be most effective; but you need a hard heart.

More considerate of the gardener's sensibilities, not to mention the plant's, is to lift the clump and split it with a little more care. This is best done soon after flowering and although it is possible, with the aid of a scalpel, to split plants into very small pieces, it is perhaps wiser for the gardener (rather than the nurseryperson) to be content with more modest aspirations and to split the plant into larger chunks, each with plenty of root. These are then potted into a mix of two parts John Innes No. 3 loam-based compost, one part sharp grit and one part sieved leaf mould (fine grade bark would be a suitable alternative). The splits are kept cool until estab-

*Hepatica transsilvanica* develops into a slightly more open plant than the more familiar *H. nobilis*

lished, then planted out, or moved steadily into larger clay pots where forms of *H. nobilis* and *H.* x *media* in particular can develop into splendid specimens.

Hepaticas are easy to raise from seed, indeed every year I find self-sown seedlings in the beds where my various colour forms of *H. nobilis* are grown. Seed is best sown fresh in summer and will germinate early the following spring. Seedlings can be pricked out at once but it is preferable to space the seed well in the seed pan and leave the seedlings in place for two years before potting; regular liquid feeding during this period is helpful.

### A singular roll-call

It is the hybrid and double-flowered hepaticas which prompt the most adulation, and as is so often the case this is in inverse proportion to the availability of the plants. A wonderful list of names is associated with these forms: Ballard, Foster, an eccentric cleric from modern times, an Irish woman, Ellen Willmott and the first royal botanist to Charles I.

The name of Ballard is mainly associated with asters and hellebores, but the most sought after of all hepaticas also has the name Ballard attached. Sadly, *Hepatica* x *media* 'Ballardii' is one of those plants for which there are far more catalogue entries, labels and proud boasts than there are true plants. First a little history.

Ernest Ballard made the cross between *H. nobilis* and *H. transsilvanica* in 1916, and although quite a few seedlings were raised, only one survived wartime neglect. This was sold first as 'Trilosa' and later as 'Ballardii' and when he showed it at the RHS in March 1938 it received the Reginald Cory Memorial Cup for the best intentional hybrid. This hybrid has the general manner of growth of *H. transsilvanica* and the flowers, although often said to be larger than either parent, are actually intermediate in size and shape; they are almost 2in across and a glowing lavender blue in colour.

Perhaps repeated division into small pieces, but more likely a virus infection, has led to this plant being available but rarely and not always easy to grow when eventually acquired; the sad truth is that strong and vigorous plants are often impostors.

More recently in the United States another hybrid turned up by chance in the Connecticut garden of Linc Foster, with the American form of *H. nobilis* as one parent. Now named 'Millstream Merlin', the flowers are smaller than those of 'Ballardii', generally more similar to *H. nobilis* in size and shape, but a rich dark purple in colour.

These two plants are very rarely offered by nurseries. However, double forms of both *H. nobilis* and *H. transsilvanica* are offered – 'less rarely' is perhaps the phrase rather than 'more frequently'. Parkinson, in 1629, describes two double forms of what we now call *H. nobilis*: *Hepatica flore purpureo multiplici fine pleno* and *Hepatica flore caeruleo pleno* were the names he gave them; thank goodness for Linnaeus. The double purple and double blue are just as rare as 'Ballardii' and 'Millstream Merlin' and only 'Rubra Plena', the double pink, is even listed by nurseries now – although if you know where to ask and are trusted as a good grower, you may be fortunate.

But Parkinson's description of the double purple sounds just captivating: 'The double Hep-atica is in all things unto the single purple kinde, saving only that the leaves are larger, and stand upon longer foot-stalkes, and that the flowers are small buttons, but very thicke of leaves [ = petals], and as double as a flower can be, like unto a double white Crowfoote before described, but not so bigge, of a deep blew or purple colour, without any threads or head in the middle, which fall away without giving any seeds.' His double blue was much the same, 'except one may say it is a little lesse in the bignesse of the flower, but not in the doublenesse of the leaves'.

'Rubra Plena' sounds slightly different, especially considering Parkinson's admittedly rather idealized illustration, with a central button of tightly packed petals surrounded by a ring of slightly larger ones, the whole flower becoming less compact before the petals drop.

Then there is 'Little Abington'. Found in a garden near Cambridge by the Reverend R. J. Blakeway-Phillips, introducer of many good plants including a true orange *Eranthis hyemalis* and a single form of *H. nobilis* in deep purplish red called 'Crawley Down', 'Little Abington' is named after the village in Cambridgeshire in which he once lived and is similar in form to 'Rubra Plena' but violet blue in colour. And going back to 1903, Ellen Willmott, no less, exhibited a double white which was awarded a First Class Certificate by the RHS.

The double form of *H. transsilvanica* has been grown for well over 100 years. The only one I have seen is the one which received an Award of Merit in 1973 when exhibited by that great Irish plantswoman Molly Sanderson, after whom the best of the black pansies was named. Shown simply as 'Flore Plena', it was then named 'Elison Spence', not 'Mrs Elison Spence', and it is slowly getting around.

### Back on earth

Returning from this guided tour of the stars, let us eschew astral travel and walk the beds in our own gardens and consider what we actually grow. For the sad fact remains that, notwithstanding Parkinson, Ballard, Foster, Willmott and the rest, most of the hepaticas grown in our gardens are single-flowered forms of *H. nobilis* in blue, pink or white. Given the right conditions they bulk up slowly into fat clumps and seed themselves about. The colour of the seedlings may vary and we can select those in favourite shades to keep and others will enjoy those we like less.

But you can always appreciate the glinting sparkle of the simple hepatica as long as you can ensure that the little stars are not clouded out by billows of other foliage; one of my best was snuffed out by an overlooked and overwhelming alchemilla seedling; fortunately it was not a 'Ballardii' which was so eclipsed.

# Bergenias

For many years, bergenias were perceived as dull, unexciting and purely functional plants with a tendency to be seen forlornly flopping around in rather bleak and featureless surroundings. Then three men, with names as unexciting as the plants themselves were supposed to be, transformed them into plants which are valued for their individual aesthetic qualities and not solely for the roles into which they can be half-heartedly pushed.

Three Smiths, and a Pugsley too, through a combination of imagination and a good eye, over the last hundred years bred a succession of new varieties which has ensured that bergenias are now taken very seriously and enjoyed with enthusiasm, rather than simply used because they can be depended upon to survive where little else will thrive.

### Ernest Schmidt

The first Smith is a Schmidt, Ernest Schmidt, a partner in the German nursery firm of Haage and Schmidt of Erfurt and whose name we know from its being attached to the early-flowering hybrid B. x *schmidtii*, a cross between B. *ciliata* and B. *crassifolia*, since 1910. This is that ubiquitous pink bergenia, passed from one cottage garden to the next, which flowers in February — inasmuch as you can see the flowers, hidden as they so often are beneath floppy foliage for which the common name of elephant's ears is something of a compliment.

In the 1960s the RHS trial of bergenias revealed that a number of similar clones of this cross were being grown, rather than just one widespread clone, and after studying them all the judges selected the best and this was named 'Ernest Schmidt', a name not seen, as far as I know, in any catalogue since.

I cannot tell you if the clone I have is 'Ernest Schmidt' or one of the less superior clones, partly because I have never seen a clone I could honestly describe as superior. I would never have grown it at all had I not inherited a 4ft clump when I moved into my previous house. And quite why it has not been used as material in the traditional art of parabolic projection in the direction of the compost heap I cannot say. But just as it begins to promise something in terms of early flowers, one of two things happens. Either those limp leaves, which never have the grace to turn a respectable winter red, droop over the flowers and hide them. Or, contrary to what is suggested by some, the flowers are melted by frost. So either the leaves protect the flowers from frost but hide them from view, or the flowers stretch up sufficiently to be seen and are then frosted. But Gertrude Jekyll liked it.

### Tom Smith

The second Smith ran Daisy Hill Nursery in Newry in Northern Ireland and along with dieramas and crocosmias, bergenias were among his special enthusiasms. He crossed B. *cordifolia* with B. *purpurascens* and the great taxonomist Engler coined the name B. x *smithii* for the resulting plants. You might think that with the name B. x *schmidtii* already in use he would have chosen something a little more distinct. Peter Yeo, who later did so much to popularize hardy geraniums,

*Bergenia stracheyi* 'Alba' was the parent of many excellent white-flowered plants

had what seemed a helpful idea in suggesting *B.* x *newryensis* as a better name but this has been frowned upon and is not in use. I refrain from further disrupting the already disordered world of horticultural nomenclature by campaigning for the retention of Peter Yeo's suggestion.

Be that as it may, Tom Smith raised and introduced at least ten varieties which were an improvement on the species available at the time and these included 'Brilliant', 'Distinction', 'Profusion' and 'Progress', names which leave us in little doubt as to his own opinions of his creations. 'Perfect', I should perhaps interpolate, was found in a garden at Düsseldorf and is not a Tom Smith plant. Towards the end of the nineteenth century when these varieties appeared they were indeed noticeable improvements. As William Robinson wrote in *The English Flower*

*Garden*: 'We have here a group of fine-leaved and flowering plants worthy of every attention, for we do not believe that the hybrids now known, fine as they are, will remain long the best now that it is seen that the species and varieties seed so freely and give such good results when crossed.' As is so often the case, although his name was on the cover it was not actually Robinson who wrote these words; but we must presume he agreed with F. W. Burbidge who did. Now, Tom Smith's varieties are only very occasionally listed in catalogues and rarely seen in gardens. Partly owing to the work of another Smith.

### Eric Smith

It took the third Smith, and another sixty years, to bring Burbidge's prophecy to pass, although the intervening years brought sufficient improvement for Peter Yeo to write at the time of that Wisley trial: 'As there is already a very good set of bergenias available for ornamental culture, it is clear that only the highest standards should be accepted in selecting new clones for introduction to commerce.' Hardly had his pen left the paper than new work was producing superb new varieties.

Eric Smith worked as propagator at Hilliers Nurseries before going into partnership with Jim Archibald to set up The Plantsmen, a nursery which soon became well known for the introduction of so many fine new plants, both wild species and hybrids. Eric worked on many plants including hellebores, kniphofias, crocosmias, and especially hostas, and from the late 1960s bergenias were one of his enthusiasms. In 1974 the listing of *Bergenia stracheyi* 'White Strain' showed he was busy.

Then in 1975, after Eric moved on, 'Brahms' was introduced, the first of eight varieties named after composers whose names began with the letter B and a selection from the white strain, later known as 'Snowblush Strain'. The outstanding 'Beethoven' followed, with neat foliage and pure white flowers in coral pink calyces, the flowers eventually blushing only very slightly – although this can vary in different soils. 'Bach' and 'Britten' were in a similar style but 'Bizet', 'Borodin' and 'Bartok' were bred from the tall crimson-flowered 'Ballawley', with its bright, glossy green foliage.

In 1977 Jim Archibald introduced the 'Winterfire Strain', seedlings originating from Eric Smith's hybridizing and specially selected for their fine winter colour. When he left Hadspen House, where he had worked with Penelope Hobhouse, Eric gave Beth Chatto a plant of this type which had been growing outside his back door and some years later she introduced it under the name 'Eric Smith'.

This is an outstanding plant, an essential for the winter garden, which when mature develops rich rolled leaves coloured in its best seasons as close to scarlet as seems possible. Neat in habit with fine upstanding foliage, this is a robust but rather slow plant which follows its superb winter colour with rich dark cerise flowers in April and May.

Many bergenias have been recommended for their winter colour, in particular *B. purpurascens*, *B. cordifolia* 'Purpurea' and 'Sunningdale', which is sometimes said to be the true form of Tom Smith's 'Brilliant'. I've also recently come across 'Wintermarchen', with small upright foliage in a wonderful red, like a cluster of crimson mayflies with their wings held high, and recent collections of *B. purpurascens* from Nepal may yield improvements though mine are not yet sufficiently mature to assess. So far 'Eric Smith' beats them all. Now, all we need is one which keeps its winter colour into spring then matches it with frost-resistant heads of pure, unfading white flowers.

### Pugsley, Bressingham and the Germans

Between the contributions of the Smiths Tom and Eric, H. C. Pugsley was at work in Derby crossing *B. ciliata* with 'Ballawley'. Unfortunately his seedlings inherited from *B. ciliata* the tendency to suffer from frost damage to the flower spikes, but the crimson 'Pugsley's Purple' flowered later than the others and avoided most frosts; it was Highly Commended in the long-running RHS trial of the 1960s. Pugsley reckoned that his reddish purple 'Margery Fish' had the largest flowers of any bergenia he grew, and it received an Award of Merit in the trial. But its flowers are often frosted.

The third in his award-winning trio was 'Bressingham Bountiful', introduced by Bressingham in 1972. Although Alan Bloom said of Pugsley's varieties that 'none were outstanding' (why then did he introduce one, and name it after his own nursery?), Pugsley himself thought 'Bressingham Bountiful' 'the most free-flowering bergenia that I grow', and 'quite a distinct colour' with its fuchsia pink flowers in brown calyces.

Bressingham themselves went on to raise three varieties, two from one sowing of open-pollinated seed saved from the white 'Silberlicht' mixed with seed from a couple of pinks. They lined out several hundred plants and selected the vigorous but rather untidy 'Bressingham White', which blushes noticeably, and 'Bressingham Salmon', which I find very slow but a good distinct, though I cannot say exciting, colour. From a later batch of open-pollinated seed 'Bressingham Ruby' was raised, the ruby applying not so much to the dark pink flowers as to the winter leaves.

This is a remarkably low-tech approach to plant breeding but does depend on having the space and the time to grow on hundreds of seedlings. But more than that it depends on the discriminating application of a good eye, for out of several hundred plants it is possible that only one or two, or even none at all, will be genuinely worthy of naming and distribution. The rest, or the whole batch, must be discarded. Plant breeders need a hard heart to go with their imagination and skill.

My plant of 'Bressingham Ruby' always retains a hint of green and the leaves, carried on long petioles, give the plant a very open and spreading habit compared with the much tighter growth of 'Eric Smith'. It looks superb surrounded by that rather neglected palest blue squill, *Scilla mischtschenkoana*, and with a spotted pink hellebore. 'Eric Smith' takes the contrast of the prolific and early, green-flowered *Helleborus cyclophyllus* and vigorous 'Atkinsii' snowdrops very well.

In recent years a succession of varieties has emerged from Germany and we find these in gardens and catalogues either under their German names, under approximate English translations or under names which appear to have been transmitted via a crackly satellite link and an ear trumpet. Hence 'Abendglut' is seen as 'Evening Glow', 'Morgenröte' as 'Morning Blush', 'Morning Red' and 'Morgan's Rod', with 'Wintermarchen' as 'Wintry Tale' and 'Winter Marches'.

I'm not sure why a semi-double bergenia is considered an advance, but 'Abendglut' claims this dubious distinction. Raised, along with 'Silberlicht' and 'Morgenröte', by Georg Arends in the late 1940s, this is a good winter variety with livery rather than beetroot red leaves. It has a sufficiently open habit for *Crocus tommasinianus* to self-sow in a mature clump and slide its slender rockets through the foliage to peep out against a foliage background of just the right shade. It depends on good growing conditions to thrive in this open way and I have *Anemone nemorosa* 'Blue Bonnet' running through mine. But in poor soil, growth is more compact and there is less opportunity for such attractive associations.

### A fortunate slice

In spite of an unusually woody rootstock, which if it grew upright would presumably lead to them being classified as shrubs, bergenias can be divided into well-rooted chunks with little difficulty. Unfortunately some, like 'Eric Smith' and 'Bressingham Salmon', are slow to develop enough material to split. But there is another way: plants can be increased by cutting up the woody rhizomes. In January or February the rhizomes are detached from the plant and all the old leaves removed. Each rhizome is then sliced into pieces about an inch long, each with an eye.

Deep seed trays are half filled with potting compost and then filled almost to the top with sand. The cut pieces of rhizome are dusted in fungicide, pushed into the sand with the compost underneath and the whole lot topped with grit. After watering in well, the tray is placed in a propagator with bottom heat in a cool greenhouse and the first shoots should appear in a fortnight, although they can be rather irregular. When well rooted they're potted up and grown on for planting out.

# Celandines and other Ranunculus

William Wordsworth, it seems, was very fond of celandines, indeed he liked them so much that they are even carved on his tombstone. But whatever his other qualities, his field botany was a little below par.

> *Pleasures newly found are sweet*
> *When they lie about our feet:*
> *February last, my heart*
> *First at sight of thee was glad;*
> *All unheard of as thou art,*
> *Thou must needs, I think have had,*
> *Celandine! and long ago,*
> *Praise of which I nothing know.*

So Wordsworth, a poet not entirely unknown for his verse on natural subjects, seems to have reached a week short of his thirty-second birthday (when this poem was written) completely oblivious to the existence of the lesser celandine, one of our most colourful early flowers. Gardeners now know it better, both as a weed and as an invaluable late winter flower.

### 'Unheard of as thou art'!

At this point in the seasonal sequence of the book it is clearly the lesser celandine, *Ranunculus ficaria*, which must engage our attention, although other ranunculi had probably best follow. At once, botany impinges again upon the world of gardeners for we have not one but four distinct subspecies under this heading, two of which grow in Britain.

The British subspecies are subsp. *ficaria* and subsp. *bulbifer*, although the many cultivated forms are derived only, it seems, from the former. In general subsp. *ficaria* is distinguished by being highly fertile but producing no mini-tubers in the leaf axils and also by a preference for more open sites. By contrast subsp. *bulbifer* produces very little fertile seed, balancing this shortcoming with clusters of tiny tubers in the leaf axils; it also tends to grow in shadier places. In spite of the fact that the former has half the number of chromosomes of the latter, hybrids are occasionally produced. These, you will be relieved to hear, are not free-seeding and free-tubering plants that grow anywhere and spread like the wind, but small-flowered plants which produce no seed and no tubers. In eastern Europe you will find the very small subsp. *calthifolius* and in southern Europe the very pale-flowered subsp. *ficariiformis*. There, all the taxonomy in one paragraph; that makes a change.

But in case your mind is firmly set against celandines, believing them to be little more than glorified weeds – what's wrong with glorified weeds? Indeed the simple wild form has a bright beauty, especially in the considerate season at which it appears. But in case you still need persuading on the subject of celandines let me quote a certain Mr Farrer, who had something of a reputation as a plantsman in his day and whose opinion is still sought in some quarters: 'all neat bright things for admittance to a cool corner … where their metallic glitter may cheer the dawn of the year'. To be fair, the dots conceal something a little less encouraging ('where nothing better is wanted'), but may I balance that by mentioning that his friend Mr Bowles also had a good opinion of them. Robinson was not a particular fan, reporting that the lesser celandine 'is so common that it would not be mentioned but for its pretty double and white varieties'.

I find them valuable not only because interesting forms of our own native wild flowers always have a special charm. But their leaves are pretty, neatly heart-shaped with or without a bold central brown streak, and sometimes speckled in silver or even completely bronzed. This foliage begins to appear in autumn and by spring makes a carpet or sometimes a mound, the leaves overlapping a little like fish scales, and around early perennials like hellebores or overwintering foliage like *Carex comans* is most valuable.

The flowers, which on our wild forms make us nervous of prolific seeding, fall into two groups of cultivated sorts – the singles and the doubles. The number of different types seems to increase every time I count them up, thirty-five at the last assessment and still growing fast. Here I shall restrict myself to the most distinctive and the most new, for in recent years, as has happened with so many plants, forms whose distinctions are less than immediately apparent (putting it kindly) have acquired names which will probably do little more than tantalize plantsmen of a historical bent in fifty years' time.

The white form is indeed pretty and while appreciated by Bowles was sneered at by Farrer: 'really only lost the gilding and gone pallid'. There are a number of white forms with different colours to the petal backs; in mine they are a slightly misty lilac and reminiscent of *Crocus chrysanthus* 'Blue Pearl'. Alan Robinson selected a delightful and strong-growing form with distinctly blue petal backs and named it after his cat, Randall. This is a doctrine of which I most heartily approve and if I ever have the temerity to name a celandine I will keep the names of my

cats and the cats of my friends much in mind. What a wonderful celandine would be *Ranunculus ficaria* 'Stripy Arnold', in fact it would be worth setting out to breed a plant specially. *Ranunculus ficaria* 'The Circus Cat' would be more problematic.

Other single celandines I grow include 'Aurantiaca' ('Cuprea' seems to be the same), in a very bold and brassy shade, and 'Major'. This is a real monster, described in the 1930–31 edition of the catalogue from Ingwersen's, who have always taken them seriously and still list a good few, as 'A handsome giant celandine; the glistening flowers are as large as a five-shilling piece, and glorious in early spring in an odd corner.' 'Major' is three or four times as large in stature, leaf and flower as our usual wild forms, though something of a spreader, and clearly demanding to be crossed with 'Brazen Hussy' to create something positively spectacular. 'Brazen Hussy' is probably grown more than any other. This is said by some to have been found by Christopher Lloyd in his garden, or by him in a wood on his farm or by his gardener across the border in Kent. In fact it was found by both, Lloyd and his gardener, while walking in a wood on the family farm.

The foliage of 'Brazen Hussy' is like polished chocolate and with the bright yellow flowers makes a stupendous combination. It self-sows and while some of the resultant seedlings may be dark-leaved, some will not, and by this means you may introduce a new weed into your garden.

Attractive hybrids of 'Brazen Hussy' have recently appeared. What sounds like one of the less exceptional has been named, illegitimately, 'Brazen Hussy Marbled Leaf', with greener patches on the dark leaf. I have not yet found this sufficiently tempting actually to buy. 'Coppernob' is like 'Brazen Hussy' with 'Cuprea'

Dark-leaved *Ranunculus ficaria* 'Brazen Hussy',
discovered and introduced by Christopher Lloyd

flowers and was found by Wendy Perry in her woodland garden at Bosvigo House in Cornwall. 'Crawshay Cream', a cream-flowered 'Brazen Hussy', was found as a seedling by Jerry Webb, who has introduced a number of good plants; the name comes from the name of the road where he lives.

I grow just four doubles but all are different in the way in which their doubling has developed and so they are worth describing. My 'Flore Pleno' (which may be 'Bowles' Double') is one of the oldest and is double in the sense that the double marsh marigold and double buttercups are double, with all the nectaries and anthers transformed into petals, which are slightly greenish on the backs. In fact dissection of the flower reveals no reproductive organs; the flower is all petals. Unfortunately they fade unattractively to white as they age. The leaves are a plain, fresh green and the plant rather bushy and reaching

10in in height. Some experts say 'Flore Pleno' is fertile; mine cannot be.

'Double Cream' is a much smaller plant with much larger flowers. Planted at the same time as 'Flore Pleno', 'Double Cream' has made a plant about a fifth of the spread and less than half the height with far fewer flowers. The flowers, though, are twice the size and the petals are pale yellow at the base fading to cream at the tips; they are shiny but with translucent streaks. The backs are cream at the base but heavily streaked in grey at the tips. The flowers have far fewer petals and retain both their male and female reproductive parts, but the feature that makes this plant so pretty is that the tips of the petals roll over slightly, presenting the tips of the grey undersides against the pale colouring of the rest of the flower. The leaves are small and dark with an occasional brown mark. This would be an absolutely essential plant if only it produced more

35

flowers; perhaps a seedling will arise one day which will be more prolific.

'Collarette', also known as 'Anemone Centred' and 'Beamish's Double', is different again. The neatest of all the doubles, each flower is like a tight button the colour of a free-range egg yolk with a ring of rounded petals setting them off. To be honest I'm rather going off this plant as I find it difficult to place in the garden, but it makes a wonderful pot plant. Finally comes 'Green Petal', with its faintly marbled leaves and its weird flowers. The petals have become largely transformed into leafy tissue and the result is a confusion of paddle-shaped organs, yellow at the base and green at the tips, with a few elongated yellow nectaries to add to the general mêlée. An oddity rather than a beauty.

I have to say that my collection of these forms has expanded greatly this last year but I will spare the tolerant reader my enthusiasm until I have grown them for a little longer.

I began by planting these varieties under shrubs, where they seemed to spread well and made an impressive, sometimes rather bumpy, carpet. But most need to be at the front of the border where their curious flowers can be inspected closely, so they then found themselves alongside hellebores and violets and snowdrops in woodsy beds, places which some gardeners might have reserved for supposedly choicer plants. Not all are suited to such places and, having seen them grown in pots, I'm coming round to the idea that the choice forms are perhaps grown better that way.

When dormant, the tiny tuberous roots which 'consisteth of slender strings, on which does hang as it were certaine graines, of the bignes of wheat comes, or bigger', as Gerard put it (or 'like so very many dainty dahlia tubers', as Rice puts it) can be split into a huge number of individual plants. For propagation purposes these can be set out singly an inch apart in deep trays to fatten up. But for potting for display, they can be kept in clusters and planted in terracotta pans. As they come into flower move them to make a display by the door, supported on upturned pots of various sizes.

### Field and creeping, fair maids and foul bachelors

Continuing this irreligious theme, let us examine the horticultural merits, such as they are, of three more weeds: the meadow buttercup, the creeping buttercup and the bulbous buttercup – *Ranunculus acris*, *Ranunculus repens* and *Ranunculus bulbosus*.

The number of nurseries which list the double meadow buttercup is extraordinary, considering how wary are gardeners of buttercups in general. But there it is, 'Flore Pleno'. It was found growing wild near Lytham St Anne's in Lancashire in the late sixteenth century, and Parkinson tells us that 'the flowers stand on many branches, much divided or separated, being not very great, but very thick and double'. I wonder if what we now grow really is a direct descendant by division over 400 years. Or whether, like many supposedly old pinks, this is a more recent almost identical replacement. Anyway, be not afraid of this dainty treasure but give it any reasonable soil that is not too dry and enjoy its commendable restraint and its delightful tight yellow flowers.

There are other forms: 'Stevenii' is larger in all its parts than the usual sort while 'Sulphureus' is pale and occasionally produces a dark patch on the foliage. Collector's items, I think – if you get my drift...

Gardeners are perhaps a great deal more justified in being cautious about *Ranunculus repens* 'Flore Pleno'. This is a rampant spreader which Gerard claims to have himself discovered while walking by the Globe theatre with a merchant friend. Although its tight, bright yellow flowers are undeniably pretty, perhaps it is, as many a writer has said of a pretty plant he cannot find a place for in his borders, 'best suited to the wild garden'. 'Any soil and position,' says one catalogue. Hmmmm. There is also 'Joe's Golden'. Now I know that *all* buttercups are supposed to be golden but this refers to the foliage, which eventually turns lime green. I do no more than lay this before you for your consideration and pass on.

Finally, the double *Ranunculus bulbosus*. Parkinson says this 'is common in every garden through England' and 'the flowers are of a greenish yellow colour, very thicke and double of leaves [his word for petals], in the middle whereof riseth up a small stalk, bearing another double flower, like to the other but smaller'. This sounds most mysterious, for this feature seems entirely absent from today's plants. This may be because what we usually grow now as *Ranunculus bulbosus* 'Pleniflorus' is usually the double form of *Ranunculus constantinopolitanus*, a wellbehaved species from south-east Europe also sometimes known as *R. speciosus*. This is a pretty plant with green-eyed flowers and good in damp spots – but nothing to do with *Ranunculus bulbosus*.

Which leaves only 'F. M. Burton', a primrose yellow form of *R. bulbosus* which I have not grown but which our friend Farrer tells us 'is not, in itself, very thrilling, but which makes an exquisite contrast if put side by side with *Aquilegia caerulea*'. I find myself curiously incurious.

I suspect there is little need to say much about propagating these buttercups. We all know about creeping buttercup, while the others can be carefully pulled apart in spring or autumn for replanting. The doubling rather prevents seed set, for the reproductive organs are transformed into petals, but it seems to me that there is no reason why more double forms of these plants should not turn up in fields today. Double and coloured celandines are now discovered every year, so why not buttercups?

Fair maids of France is a name which has been applied to three different flowers. Quite why it came to be attached to the dog daisy, *Leucanthemum vulgare*, I cannot say. Neither is it entirely clear why the name came to be attached to the meadow saxifrage, *Saxifraga granulata*. Turner in his sixteenth-century herbal reports that the 'little knoppes lyke pearles', the overwintering buds, were sold as *Semen Saxifragae Albae* and Gerard mentions that boiled in wine they were 'singular against the strangurie and all other griefs and imperfections in the raines'. Perhaps, when these problems were solved, the more licentious 'gentlemen' of the time could again turn their attention to the fair maids of France, being notoriously more open to enticement than English women. This saxifrage was also called pretty maids, hence the nursery rhyme's pretty maids all in a row. It has been mischievously suggested that once cured of 'imperfections in the raines', a complete row of maids was necessary compensation for painful abstinence. But being a supposedly reconstructed sort of a chap, I dismiss such speculation with a lofty sneer – and repeat it solely in the cause of entertainment.

Be that as it may, the plant I am actually coming round to discussing (after so much wanton conjecture) is *Ranunculus aconitifolius* 'Flore Pleno'. One of the great delights of the cottage garden, this far from flamboyant plant is all the better for its cool restraint. The contrast between the dark green, dissected foliage and the exquisite simplicity of the neat white double flowers gives this plant a very special appeal.

Oddly enough, it is also occasionally known as fair maids of Kent (I can't think why) and for many years languished in sad obscurity. Now that its qualities have been rediscovered the single form is also occasionally listed, though I have not yet seen it. The double prefers a dampish soil where it will reach 3ft and make a prolific clump; in drier conditions it will make 15in and languish. It can be split, with a stout knife, in the autumn. We're perhaps only recently discovering the advantages of plants like the fair maids which disappear promptly after flowering rather than slowly and inelegantly hanging on. Of course it means we must take care where we dig, but later perennials or climbers can be trained over the spot or we can sow some quick annuals like white nigellas for later.

It is perhaps worth mentioning bachelor's buttons in this context. A bachelor's button was made up of a series of small squares of cloth laid one on the other and secured to the jacket with a single stitch. It signified that the wearer was unmarried and unattached, and these eager bachelors were also said to carry flowers that resembled their bachelor's buttons to present to their chosen maids. So fully double flowers like these double ranunculus came by the name.

## The Asiatics

In these strange days when capitalism has been allowed its head and at times looks like shaking it so hard it will fly off, multinational business has had a selective and unpredictable effect on the choice available in so many products giving us an apparently endless selection of hair conditioners and impatiens but really only one draught stout and two florist's ranunculus.

For these days there seem to be just two types of ranunculus that take up so much of the market: 'Accolade' and 'Bloomingdale' for growing as pot plants and 'Victoria' for cut flowers; all three are $F_1$ hybrids to be raised from seed. It's perfectly true that mixed 'claws' are available for cutting

(they're usually called 'Mixed') and the results are mixed too. But where are all those old ranunculus? We do still grow some single forms of *Ranunculus asiaticus*, direct descendants of wild forms, but in spite of their simplicity, their wonderful purity of colour and their indescribable sheen they are rarely seen.

In the eighteenth and nineteenth centuries things were very different. One catalogue, published in 1847, gives 145 different varieties, and Sacheverell Sitwell made a selection on his own terms: 'We extract from this list, for their poetical ring, the varieties named Faunus; Menelaus; Curion; Passe Favourite Mignon; Manteau Imperial; Roi de Mauritaine; Zebulon, yellow striped; Aglaia; Feu Eclatante; or Bizard Singulière. And this, we must remember, was in the decline of the Ranunculus. Its apogee could be placed about 1760 to 1770.' James Maddocks, who had a nursery at Walworth, now a London suburb, grew about 800 different sorts. Today we have little more than a strain named after an American department store.

These old florist's forms with their delicate picotees, stripes and speckles need not be lost for ever, and indeed are not, quite. Timothy Clark in Cambridgeshire keeps them going, sowing his own seed every year, and he tells me that the virus which causes so many of the pretty patterns is the same that causes breaking in old florist's tulips, which he also grows.

It was Janus, was it not, who looked both ways at once? Well, we seem to live in an age in which we strive fiercely for the new and for discovery for its own sake; and this is how it must be, always provided that we have the wit to realize which of our discoveries are valuable and which are not. But at the same time we turn and look longingly back to the past in a nostalgic haze, even to the extent of trying to recreate its flowers as much as its moral standards. But I wonder if the flowers of the nineteenth century were really as exciting as its morality was said to be rigid. Memory and historical assessment can be very selective, and while we look at idealistic paintings and engravings and long for those plants, portrayed as more perfect than many ever were in life, so we read the guidance of moralists and clerics and assume that the world they wished for and encouraged people to create also actually existed. I suspect that both the flowers and the morals were a great deal less perfect than we are led to believe.

# Winter into Spring: A Final Choice

This period of winter into spring is one in which exploring the emerging garden is like searching for coins down the back of the sofa; you know they're in there somewhere but at first you can't always find them. Eventually you'll find the coin, you know you will, and as you inspect the gaps in your beds and borders, swaddled in jumpers, you know that there too you will find them in the end, be they the insidious green spikes of couch grass or the more rewarding glitter of cyclamen.

### What do you mean: 'It's an alpine'?

Even twenty years ago, if I'd written that cyclamen are essential plants for the winter garden, people would have said: 'I thought they were pot plants for the windowsill.' Now, many gardeners look down on the pot plants in spite of recent breeding developments which have led to daintier flowers, prettily marked foliage and even the reappearance of a little scent, preferring instead to grow the hardy species which few gardeners even knew about until relatively recently. True, Ingwersen's, the alpine specialists, listed eight in their 1930–31 catalogue and they were grown by a few specialists. In the 1990s they are on sale in supermarkets alongside the flamboyant pot plants, and the ones I shall choose at least do not demand alpine house treatment.

So it's now perfectly true to say that the hardy cyclamen is vital for winter flowers, and much of this rise to indispensability is the result of work by the Cyclamen Society. Not that they go round badgering the chain stores into putting hardy cyclamen on their shelves, but their research, both horticultural and botanical, is leading to a very sophisticated understanding of the distribution and habits of cyclamen in the wild, their taxonomy and their cultivation in gardens, so that gardeners now have a far greater awareness of what it is they are growing and how best to grow it. And the number of species and forms available from nurseries is growing all the time.

The foremost species for the winter garden is *C. coum*, which is being studied in depth by the Cyclamen Society, whose field trips to Turkey and Israel have added enormously to our understanding of the species. But it has needed little research to reveal that in the right conditions in the garden, this is a wonderfully dependable plant. It needs good drainage and partial shade, and a leafy, woodsy soil suits it well. Its preference for these conditions enables it to associate well with many other winter flowers like hellebores and *Crocus tommasinianus*, and its tendency to self-sow greatly adds to its appeal. In the wild it sometimes grows with *Helleborus orientalis* and *Primula vulgaris* subsp. *sibthorpii*, an association which should prompt ideas on other garden companions.

The plants are small, with neat rounded foliage which may be dark green or marked in a number of ways. I like the richness of the plain dark leaf, especially in its white-flowered form, but there's no doubt that those with silver patterning are more valuable when not in flower. Some of the more choice of these, like 'Nymans' and 'Tilebarn Elizabeth', are especially beautiful but not always so robust in the garden as the green-leaved types.

They are also hard to find in nurseries, but some sell unnamed silver, pewter or patterned forms and these are ideal as starts for the garden.

The flowers of *C. coum* are appealingly dumpy, and as with most species each petal has something of twist to it. They come in various magenta and pink shades, all with a darker mark at the base of the petal, and two different white forms. One has the same dark mark at the base and is quite common and easy; the other, 'Tilebarn Fenella', is a relatively recent introduction in pure white with no basal mark, but is too rare, and too expensive, to take a chance with outside in the open ground – always assuming someone will sell you a plant in the first place.

*Cyclamen coum* is the most dependable of the winter cyclamen, but for milder areas *C. libanoticum* is worth trying. I say 'milder areas', but as more people have been growing this species they are finding it hardier than was first sus-pected. This is quite a bold plant, with prettily patterned leaves in green and blue-grey. The flowers are unusually large, opening almost white and then blushing as they age. Good drainage is again vital and soggy soil in winter is a killer. Well-drained woodland conditions are ideal, perhaps in a made-up bed, but even when perfectly content self-sown seedlings are uncommon.

The next in this winter quartet of cyclamen is *C. pseudibericum*, which is also a great deal less reliable outside than *C. coum*. The foliage is dark green with grey-green marbling and the large, rather striking flowers are usually magenta with a purplish mark at the base. Unlike *C. libanoticum*, which is less pleasantly scented, the flowers are quite sweet.

Finally, *C. trochopteranthum* with its flowers like whirling propellers. And what a name – though it does mean, well, 'with flowers like whirling propellers'. The leaves are prettily

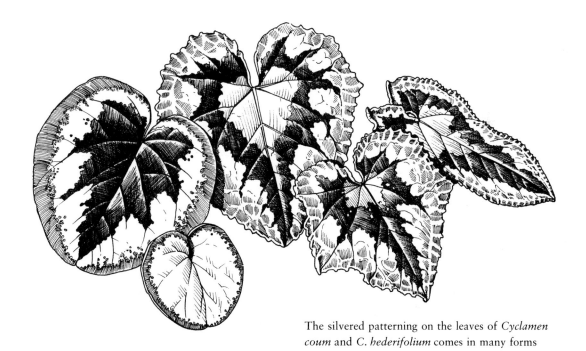

The silvered patterning on the leaves of *Cyclamen coum* and *C. hederifolium* comes in many forms

marbled and the petals, instead of being completely reflexed, would seem only half reflexed were it not for the sharp twist in each which gives it its propeller-like look. This again is one with which you take something of a chance.

But perhaps it would be fairer to nominate C. *hederifolium* as second choice to C. *coum* for winter. It would not be unreasonable for hesitant gardeners to resist taking a chance on these other species, as they are not the easiest to grow, a reservation that could not be applied to C. *hederifolium*. Of course C. *hederifolium* flowers in late summer and autumn so in winter we grow it for its foliage.

The range of its leaf patterning is simply wonderful and I especially like those with leaves shaped like arrowheads and with silver patterns following the general outline of the leaf. There are other silvered sorts, and one named 'Apollo', found by Gerard Parker among Bowles's plants at Myddelton House, is especially attractive, but, as with C. *coum*, rather than risk the rarities start with unnamed silver seedlings in white and in pink and allow them to self-sow.

The leaves retain their freshness right through the winter and into early summer, and being so adaptable in the garden can be planted in a variety of situations. In fact self-sown seedlings will usually appear unasked in sufficiently unexpected places for their adaptability to be convincing. Start by planting in well-drained woodsy conditions in dappled shade and let them find their own homes from there.

I have a contented and expanding plant nestling among two clumps of *Carex comans* 'Bronze Form', and with the sedge cut back in midsummer to ensure the leaves are not too untidy in winter it makes an intriguing picture. And as the nearby 'Blue Bonnet' anemones open in spring the silver, the milk chocolate of the so-called bronze sedge and the misty blue are just right.

## We few, we happy few

If we exclude the true alpines as we presumably must, we cannot but conclude that winter is a scarce season for perennials. Hellebores, bergenias and cyclamen I have discussed, hepaticas, celandines and winter iris too. Little remains unless we borrow from spring and autumn, which I refuse to allow – or at least only a little.

*Primula megasifolia* is a true winter perennial, although something of an unknown to most of us. It flourishes in southern woodland gardens, where the trees provide frost protection both for its umbels of early pink flowers and its bold evergreen foliage – not to mention vital summer shade. *Megasea* is an old name for bergenia, which tells you everything you need to know about the appearance of the foliage. The intriguing hybrid 'John Fielding', with *P. juliae* as the other parent, is hardier and easier to grow; my attempts to cross 'John Fielding' with various wild and cultivated primroses have led to ignominious failure.

In Britain we have a sweet-smelling weed which only gardeners with a self-destructive streak introduce into their gardens. It is a joy to see and to smell, after parking in a roadside lay-by and walking along the verge to inspect it. But beware – *Petasites fragrans*, the winter heliotrope, is more an armed invasion than a garden plant and we would enjoy its heady, almost sickly fragrance so much more if only it were not such a thug.

Another, much larger and even more alarming species, *P. japonicus* var. *giganteus*, is planted in gardens more often, for its huge, floppy, 4ft leaves. Unless you have a grand estate and can plant it so far from the garden that you need to go by Land Rover to inspect it, eschew this plant with fervour. If you have it so near that you need do no more than stroll down the garden path in your slippers to smell it, in a couple of years' time it will be bursting up through your kitchen floor and toppling the cooker. But we are discussing early-flowering plants, and this one blooms in March so should perhaps be excluded from this chapter as well as from the garden.

In recent years Joe Sharman of Monksilver Nursery has developed a characteristic enthusiasm for other *Petasites* species, some of which he has collected in the wild and catalogued for the benefit of the seriously deranged. 'Creeps slowly to form a dense clump,' he says about one species, not previously in cultivation. Do you really believe that? Only gardeners with a fundamental and perhaps unacknowledged streak of masochism would plant such beasts anywhere but on the bonfire.

### On loan

Finally two plants which we can surely borrow from spring, including one related to the British native lady's smock, *Cardamine pratensis*. There are a number of delicate and delightful spring flowers in this group, many now dispatched to the genus *Dentaria* (and promptly returned by some botanists), but the only one we can legitimately borrow from later months is *Pachyphragma macrophyllum*, formerly known as *Cardamine asarifolia*.

Unfortunately this is the least refined of the group, 'coarse and unworthy', says Farrer; the bright green leaves are more reminiscent of a hedgerow weed than a cherished woodlander and the pure white flowers are large and straightforward. But it is certainly early-flowering, certainly bold, and the purest, cleanest of whites set against its green foliage. We should appreciate it for all that and give it a place in full sun or partial shade where it will never dry out.

Neither could we borrow most pulmonarias from later in the year; only *P. rubra* belongs here.

I have often seen this opening its first coral red bells in January and this year, like Bowles's often did, mine flowered in December. At such a season, with snowdrops scattered round and an early hellebore opening overhead, how can we not welcome it? 'Redstart', we know, is simply a name coined by Jim Archibald when the RHS insisted he distinguish his entry for the Wisley pulmonaria trial from another stock entered as *P. rubra*; in this case it is not one of the many special selections made by The Plantsmen nursery.

But 'Bowles' Red' is a particularly good form with a few faint spots on the foliage, 'Barfield Pink', with curious white stripes on the flowers, is another good one, although the newer 'Ann' looks better, and recently Richard Nutt, who gardens at Great Barfield in Buckinghamshire, has introduced 'Barfield Ruby', said to have the largest flowers of all. 'David Ward', found in Beth Chatto's garden by her propagator of that name, has a narrow cream edge to the leaves; this looks dramatic indeed, but may scorch. Perhaps most sought after by those who have had news of its existence is a rare white-flowered form, 'Albocorollata', which, I confess, I have only recently acquired after years of covetousness.

Finally, to look back to lingering autumn, *Aster tradescantii* hangs on like a cross between an asparagus fern and a daisy, but so valuable in being fresh-looking, reliable and self-supporting. *Saxifraga fortunei*, such an exquisite plant with its liverish leaves, especially in the shining crimson 'Wada's Variety', may also linger through those autumns where frost touches but lightly. I always enjoy those five-pointed flowers, like marionettes swinging their legs from crimson wires, and on the leeward side of a daphne, sheltered from frost and gale, they should last long enough for inclusion.

# SPRING WOODLANDERS

# SPRING WOODLANDERS

Is it unfair to point out that in Gertrude Jekyll's *Wall, Water and Woodland Gardens* a mere eighteen pages out of over 450 are devoted to woodland gardens; and some of this is taken up enthusing about rhododendrons? Not that there is anything wrong with rhododendrons, I suppose, in the right place – which as far as I'm concerned is usually in somebody else's garden. She also, by the way, suggests growing cistus among the azaleas, an uncommon notion which I shall not be taking up.

My own enthusiasm for spring woodlanders concerns plants of a different class, plants which would recoil from the glare and garishness of many rhododendrons and azaleas. Fortunately I garden on lime, so temptations to succumb even to *R. cinnabarinum* or *R. glaucophyllum*, which would certainly be bearable, simply do not arise. A rhododendron and azalea garden in full cry really is a riot of colour and I can do without a riot in my back garden.

No, these are perennials of delicious, deciduous woodlands the temperate world over. Delicate and often conspicuously dependent on the conditions they enjoy, these are the plants which root into the leafy litter for their early moisture and nourishment, and their growth begins almost before winter has ended so that their new foliage can soak up the straining spring sun before the canopy cuts it out. By remaining evergreen, a few take every weak winter ray. In summer they may retreat into dormancy as the trees leaf out and transpire so effusively, their roots extracting comparable moisture from both leaf litter and subsoil. They flower as the first insects are abroad; their seeds and seedlings are adapted to the continuing dark-warm-and-dry/bright-cold-and-wet seasonal regime. We only need to look at the woodlands in which they grow to understand the conditions they appreciate in the garden.

Occasionally, we cheat. Some hostas, for example, grow naturally not in woodland but in entirely open situations, though often supplied with liberal ground water to counterbalance the lack of shade. It can work both ways, and this is where the skill of the gardener rather than simply the observer or researcher is crucial.

But, as any perceptive reader will probably have guessed, this is little more than an excuse to enthuse about all those captivating spring shade-lovers – and a few of their sun-loving friends and relations. And captivating they are, though a little demanding too. 'What a blessed time it is for garden and gardener when the wind goes round to the south-west and warm April showers begin to fall,' says Mr Bowles. The plants enjoy the rain, and we enjoy the plants.

47

# Arums

We British can be very obtuse at times. It must be the way in which so many of us are brought up, encouraged to be polite, civilized, demure and cautiously friendly, even when we feel angry, hostile, licentious or downright bloody-minded. We have sayings for it: 'Children should be seen and not heard', 'Sit quietly and eat your tea', 'Least said, soonest mended', the truly debilitating 'Ask no questions and we'll tell you no lies'. What else can explain the prudish attitude to one of our most distinctive wild flowers?

## Fine and upstanding

Our cuckoo pint, *Arum maculatum*, has the usual arrangement found in arum flowers. Cuckoo pint is the very oldest of its common names and is derived from the Anglo-Saxon 'cucu' meaning lively and 'pintle' or pint meaning penis. Robert Turner, writing in 1548, spelt the name cuco-pintell. The reason for this could hardly be more obvious.

From the days when Turner asserted that 'God hath imprinted upon the Plants, Herbs, and Flowers, as it were in Hieroglyphicks, the very signature of their vertues', the blatant appearance of this flower prompted its use as an aphrodisiac. Along with these ideas goes an enormous string of male and female names. Lords and ladies is now the most familiar but, to select just a few, bulls and cows, kings and queens, stallions and mares, sweethearts, and Adam and Eve have also been heard.

What is so startling is that an entirely different and altogether more quaint and, frankly, rid-iculous piece of reasoning is also given for these names. *Arum maculatum* comes in two forms, one with a purple spadix (the 'pintle') and another, much less common, with a yellow spadix. 'Since purple is a lordly colour,' writes C. T. Prime in the late 1950s, 'and yellow a feminine one, what is more natural than the name lords-and-ladies?' The prurience and quaint disingenuousness which insists on such a coy explanation for so obvious a sexual image is also, perhaps, behind that extraordinary old English proverb, which has now been thankfully dis-credited, 'Say no to pleasure: "Gentle Eve, I will none of your apple."'

And we cannot stop there. For not only does 'cucu' mean lively but, to quote Geoffrey Grigson: 'There seems some folk-etymological confusion between cuckoo, cuckold and cuculle = hood, cowl, Latin cucullus, in the name Cuckoopint: here's the penis of the lecherous cuckoo, who cuckolds the birds, and here's the penis inside a hood, or in a cowl – that is to say, the penis of the priest, or monk, or friar who goes round cuckolding husbands, like the man who reads the gas meter.'

Perhaps most strange of all is another name quoted by Prime, 'Kitty-come-down-the-lane, jump-up-and-kiss-me', which seems to confuse the issue; should it not be 'Kitty-come-down-the-lane, *I'll*-jump-up-and-kiss-*you*'? Or are we now dipping our toes into uncomfortably deep water? And you thought this was a book about gardening...

*Back in the garden*

As it happens, we don't grow *A. maculatum* much in gardens. A vigorous seedling has turned up in the middle of my hot and dry Mediterranean planting, which was something of a surprise, and in some woodland gardens it can even be a weed, but generally *Arum italicum* is the one we grow. As a species, this too is a British plant but the subspecies from which most cultivated forms derive grows in mainland Europe and not in the UK.

The form with the striking white veins to the leaves, subsp. *italicum*, is a superb garden plant and unlike *A. maculatum* the leaves emerge in the autumn, making this an invaluable winter foliage plant. None of this explains why you find it placed in this particular chapter – it flowers in spring. Its scarlet fruits, maturing in September, are also a special attraction. Bowles tucked the tubers between the rhizomes in his iris beds and enjoyed their brilliance among the iris foliage in autumn.

There has been some confusion – how many times will I write something like that when I come to discuss the names of a particular group of perennials? It really is becoming very trying. Botanists around the world work on the same group at the same time and come to different conclusions; work from the former Soviet Union or India or China comes to light which changes the accepted way in which we look at things; gardeners apply names to plants with little or no regard for the quality of the plant so named, whether it already has a name or even its similarity to existing varieties. If only...

As I was saying, there has been some confusion about the names of these white-veined arums. *Arum italicum* subsp. *italicum* is the correct name for the white-veined plant which grows wild over much of southern and western Europe, though not in Britain except occasionally as an escape from gardens. Two other names, subsp. *marmoratum* and subsp. *pictum*, are simply synonyms for this plant. (*A. pictum* is a distinct, less hardy, white-veined species from the Balearics, Corsica and Sardinia.) Here it is perhaps worth mentioning that in Britain we have a different subspecies of *A. italicum*, subsp. *neglectum*, which is confined to south and south-western areas and is clearly distinguished from *A. maculatum* by its winter-green foliage. The veins in its leaves are just a little paler than the leaf blades.

While subsp. *italicum* and its forms are the most frequently grown of all arums, there is one other interesting subspecies which is occasionally found, subsp. *albispathum*. This is much like the British subsp. *neglectum* except that the inside of the spathe is pure white and the outside green; interesting rather than essential.

Apart from one or two still grown under collector's numbers there are four named forms of subsp. *italicum* found in gardens. 'Chameleon' was selected by Henry Ross, Director of the Gardenview Horticultural Park in Strongsville, Ohio, and is described by Tony Hall, under whose care it was grown at Kew, as 'like cooked spinach into which thick cream had been stirred'; or more formally, 'marbled grey-green to slate green with yellow veins and a dark green margin'. 'Nancy Lindsay' is a monster form reaching 2ft in height whose leaves emerge yellow in early winter and with narrow cream veins. 'Tiny' is simply a small form reaching about 9in with black blotches on the foliage as well as white veins. 'Taff's Variety' is the same as 'White Winter', which in my garden at least has reached over $2\frac{1}{2}$ft in leaf with some leaves as long as 15in, although it is generally said to be on the small

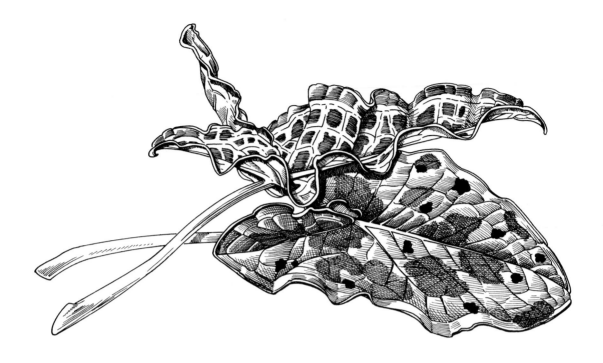

Young leaves of *Arum italicum* 'White Winter' (*top*)
and the hybrid 'Streaked Spectre'

side. It confidently muscles aside the self-sown seedlings, both red and green, of *Tellima grandiflora* 'Perky'.

'Tiny' I grow with the double bloodroot, *Sanguinaria canadensis* 'Multiplex', while *Fritillaria pyrenaica* slowly clumps up behind – another clump of the frit thrives more heartily in full sun; a good, dark, self-sown *Euphorbia amygdaloides* 'Rubra' completes the picture.

Hybrids are not common, although in south and south-west England and the Channel Islands sterile hybrids between *A. maculatum* and *A. italicum* subsp. *neglectum* are occasionally found, while at the University of Washington some hybrid arums arose in a mixed planting from which Jerry Flintoff selected a couple of es-

pecially good forms and named them 'Streaked Spectre' and 'Green Marble'. Both are vigorous and bold in leaf and make superb winter foliage plants.

And so back to *A. maculatum*, the subject of C. T. Prime's fascinating though rather prim monograph. Although specifically cultivated in few gardens this species is a useful wild garden plant, as much for its brilliant orange berries as for its foliage or flowers. In limestone country it tends to appear spontaneously in garden hedges and shady borders and the clumps of stumpy orange fruiting stalks make quite a feature, although the plants can be invasive. It might also be worth making selections from wild *A. maculatum*, choosing those forms with leaves especially strongly blotched and puckered in black.

Of course, we grow other arums, most of which are best in a gritty raised bed or even the alpine house rather than the woodsy conditions needed for the species discussed so far. The spectacular *A. creticum* is a great treat, although I've found my seedling heartbreakingly slow to reach flowering size. The fresh, glossy foliage emerges in autumn, and if lifted above ground in a raised bed seems to be given protection enough from the worst of the frost, a technique which also works at Kew for *Helleborus vesicarius*. A word of warning, though: frost-hardiness in this species is partly dependent on where exactly the stock originates, high altitude plants naturally being hardier.

This species has two impressive claims to fame. One is that it is sweetly scented (a feature which disqualifies it from an appearance in the next section), the other is its pure yellow colour. I also like the way the flower sometimes flops over on its stem and the yellow spadix pokes out so boldly. In the wild, forms with a white spathe and red spadix occur and provoke singularly irreligious covetousness; those original Velvet Underground records you've treasured since 1969 would perhaps make a realistic swap.

One more... The sight of a clearing on a Greek island dotted with *A. dioscoridis* is enough to give anyone the giggles. Those huge spathes, deep purple at the base then breaking up into blotches and that long wiggly purple spadix and the smell, urgh, and the sheer size of the whole thing... repellent and captivating at the same time and so, of course, indispensable. As it happens this one is hardier than many Mediterranean species and not difficult to grow, but leafing as it does in the autumn, frost pockets are certainly best avoided.

### And their friends and relations...

Five genera complete this group of hardy aroids: one highly fashionable genus, two very much in the second division as far as most gardeners are concerned, one which is most definitely more intriguing than spectacular, and one which is decidedly obscure.

When a particular genus begins to find favour with enthusiasts and collectors a number of things happen. First of all plants start appearing in nursery catalogues at inflated prices. In many cases, with both demand and prices high, plants are collected from the wild to meet this demand, especially if the coveted species are slow or difficult to propagate. Then, as less expert and less responsible nurseries start to cash in, names become muddled, common species are passed off (*sometimes* accidentally) as more sought-after species, variants are glorified by the assertion of their hybrid origin and unremarkable forms dignified by cultivar names. Then a monograph appears which sorts everything out to the satisfaction of the majority of growers but sets a few fanatics at each other's throats as they argue arcane points of taxonomy and provenance.

Most of these embarrassments have befallen arisaemas in recent years, and only the application of superfluous cultivar names seems to have passed them by; but then, they are so slow to propagate that hybridization is virtually unknown and propagation is almost entirely vegetative, so variation is less likely. But it will come, albeit in moderation. The monograph is certainly on the way.

To gardeners the arisaemas are distinguished from the arums by their divided leaves; in general they are also more elegant and in a few cases more colourful. Some are rendered especially intriguing by the presence of long tails on the tips of the spathe, spadix and occasionally on the tips of the leaflets. In one, at present known as *A. thunbergii* subsp. *urashima* but which has at times been segregated into a genus of its own, the tail on the spadix can reach 2ft in length and, as the plant is quite short, this lies out flat on the ground. There are around 150 species in all, with forty, perhaps, in Japan alone; fortunately only a few seem to have been the subject of nomenclatural meanderings.

*Arisaema atrorubens* is interesting for its name as much as anything. After all that prissy pretence of the British names, this most widespread of American arums in the eastern states is widely known as Jack-in-the-pulpit and also by the Quebecois equivalent, petit prêcheur, which does not seem to avoid entirely a slight sexual connotation.

Oddly enough, there is rather a sad connection between *A. atrorubens* and the *Rosa canina* of the English hedgerow. Like other arisaemas, this species develops a substantial starchy tuber and is so common in some woodlands that it would seem a possible source of food. But even prolonged boiling does not remove the poisonous alkaloids; so the tubers remain inedible. The arrogant settlers gave it the name Indian turnip in rather the way that the vigorous but fleeting single-flowered rose of the hedgerow was christened the dog rose because it was reckoned so inferior to garden roses. Oignon sauvage is but the Quebecois equivalent.

By far the most popular of the arisaemas is *A. candidissimum*, which is hardly surprising considering the attractive pink and white stripes on its hood and its tolerance of more open situations, and more open soils, than other species. *A. flavum* is altogether less flamboyant, with squat little flowers in a variety of colours, but is very distinct in that the tip of the spathe is folded right over the spadix so that the point hangs down the front creating an effect, as Linc Foster put it in his valuable summary of the arisaemas, 'like a small blinking owl'.

Another of the more colourful species, 'the most glorious of the genus' according to Linc Foster, is *A. sikokianum*. It is modest in height, the leaves are patterned in silver and the spadix is white with a disconcerting swelling at the tip. The spathe is rich bishop purple outside and pure white within and stands up erect, but as it ages it softens and becomes limp while the purple stains seep through from the outside.

*Arisaema flavum* usually has both male and female flowers on the same flower spike, but most species are very different and can even change sex during their life. As a rule, young, slow-growing, unthrifty and generally rather feeble plants behave as males, while more mature and more vigorous plants are female!

The first time I took a walk along the coast on the first evening of my first visit to Majorca I found two plants that I did not know, growing together. I knew *Arisarum proboscoideum* from many a woodland garden, where you part the leaves to find the posse of mice, tails up, disappearing earthwards. But I had never seen *A. vulgare*, a delightful little plant of short turf near the sea where it grew with a dainty, pale blue daisy, *Bellium bellidioides*. Clearly a less hardy species, requiring brighter and better-drained conditions than *A. proboscoideum*; a gem for that Mediterranean house that I keep promising myself and for warm gardens.

Lysichitons fall into the second division of aroids and most gardeners prefer to enjoy them in other people's gardens. The foot-high, buttercup yellow spathes of *L. americanum* are the more familiar, and in Britain have even become naturalized on brooksides downstream of bog gardens where they spread happily in damp conditions. Seeing them growing wild in soggy woods in Washington State was an eerie experience, the only flowers in the wood in early spring. But splashing through the rotting branches and the black ooze, it was clear that not all the plants were quite the same shade of yellow; garden forms all seem the same.

The white *L. camtschatcensis* is much less often seen but has an appealing air of cool superiority. Hybrids are creamy in colour but rarely found. I suppose if you grew a plant of each in large pots, placed them side by side under an insect-proof net and introduced a swarm of flies, cross-pollination would be bound to take place. Or perhaps they're pollinated by some sort of water beetle.

Pinellias may not be showy, but they hardly deserve the harsh words of Jellito and Schacht, 'these are not particularly attractive plants', for *P. ternata* is a delightful, if obscure, little plant, making an elegant stand of short stems with green cowled flowers. Perhaps the feature which is so decisive in confirming the attractiveness of all these aroids is that they do seem to have a personality, it's easy to – well, anthropomorphize is not quite the word – to treat them almost as pets, or, in the case of many species, rather badly-behaved friends.

Finally the smell to end all smells, *Symplocarpus foetidus*, a weird plant indeed which flowers in February or even January and whose flowers are like colonies of red speckled toads sitting squat in the soggy bog garden just where you'd expect to find them. They are pollinated by ground beetles, and anyone fool enough to get down on their hands and knees in a bog in February deserves to get a foetid noseful. Perhaps this is worth growing as a specimen in a huge pot standing in a dish of water. Yes!

## Seeds and splits

Raising true arums from seed is not difficult but it can be slow and also a little mystifying for the uninitiated. Arums fall into two groups. One group, which includes *A. maculatum*, grows naturally where the winters are cold and germination takes place with absolutely no sign of foliage above ground at all. Instead, roots and a tuber are produced below ground during the first year and leaves appear the year after. In the other group, which includes *A. dioscoridis*, whose members grow in areas with warmer winters, germinating seeds produce a rudimentary leaf in their first season as well as a tuber.

Two other things need considering. First of all it pays to wash all the skin and pulp away from the seed before sowing, to remove germination inhibitors and prevent rots which attack the pulp, moving on to attack the germinating seed. Secondly, germination usually occurs at about the time of year that mature plants of the same species are producing their foliage. Although of course if this is not made clear by the appearance of an actual leaf, the hapless gardener is left in disconcerted ignorance. Especially as old dried seed may miss a whole year before doing anything.

In effect it all boils down to this. Sow the washed seeds in 5in pots filled with a compost which is limy, humus-rich, yet well drained. Cover the seeds with their own depth of compost, top off with $\frac{1}{4}$in of grit and leave them in a cold frame, never allowing them to dry out, until something happens. When it does, feed with a half strength liquid feed at every watering. When the top growth dies down the young plants can be potted individually and eventually planted out. Seedlings may take five years to flower, although *A. dioscoridis* may flower in three.

Arums can also be divided, and while most

will stand division at almost any time, disturbing them in growth sets them back severely. The start of the dormant season is perhaps best for plants in the garden as they will be easy to locate from the remains of the top growth.

Arisaemas can be slower from seed, but germination at least can be speedy if the seed is sown fresh. When dividing it's often wise to detach small side growths from the tuber while they are still small, rather than attempt to split off more mature tubers. Lysichitums can be split, and raised from fresh seed, but symplocarpus have a fat vertical tuber which is difficult to split and so seed, again sown fresh, is a more reliable method of propagation.

*Opposite*
The toothed evergreen foliage of *Helleborus argutifolius* looks effective in the January frost, yet even then the first green cups are starting to open – a surprise, perhaps, for a native of Corsica.

*Overleaf (left)*
ABOVE On a frosty, early spring morning the red-tinted bracts of *Euphorbia polychroma* 'Purpurea' unfurl to reveal the yellow buds, already starting to show their distinctive colour.

BELOW Six weeks later the plant has expanded into an unrivalled dome of colour; an intermingling of rich blue *Scilla siberica* around the edge would complete the picture.

*Overleaf (right)*
ABOVE The bold, silver-patterned foliage of *Pulmonaria* 'Roy Davidson' follows its neat, pale blue flowers and makes an ideal background for stray strands of *Clematis* 'Hagley Hybrid' in the Withey/Price garden in Seattle.

BELOW Wispy pale green strands of *Carex buchananii* 'Viridis' fall through the distinctive steely grey-blue foliage of *Euphorbia nicaeënsis* to make an intriguing combination.

# Hostas

As Christopher Lloyd posed the question, what can be said about hostas that has not already been said many times over? Well, something, clearly, or the chapter would end here and there seems plenty still to come. But I shall take to heart the remarks that close his passage on hostas in *Foliage Plants* (the previous question opens it): 'It would be as easy to write too much about hostas here as to grow too many of them in the garden.'

In fact, I think I might just be safe from that particular mistake, as my own collection of hostas is but a modest one – I was going to say that I have been rigorous in my choice but as some seem to have chosen me rather than I them, I cannot deny that I have ended up with rather an odd collection.

But there is someone who clearly sees Christopher's advice as absurdly restrictive, and that is Wolfram Georg Schmid, author of over 400 large, densely printed pages which comprise *The Genus Hosta*. This is a triumph of a book which, although it tells me just about all I want to know,

also tells me far, far more than I shall ever have cause to inquire about.

But having mentioned Mr Wolfram Schmid and his book for the gourmand, I cannot pass without a word on Ms Diana Grenfell. Her more selective presentation in her book *Hosta: The Flowering Foliage Plant* clearly has the gourmet in mind, being a great deal less likely to upset the stomach of the everyday gardener, who is naturally more interested in a light and varied diet than in stuffing himself silly.

So now I must boil down this dilute liquor of confusing and sometimes confused detail and present you with one carefully prepared meal, complete with a few piquant touches. It sounds impossible to me, but here goes: a few hostas make good flowering plants; many hostas make good foliage plants, some are superb; most are very easy to grow; there are far too many varieties; slugs love them. I think that contains the basic nourishment although it's not exactly spicy.

## First course

Three of my most effective hostas I grow more for their flowers than for their foliage. One is *Hosta* species AGSJ 424, which hardly anyone grows, but we'll return to that in a moment; 'Hadspen Heron' is another, which is listed by an increasing range of nurseries; and finally comes 'Honeybells', which everyone grows. And so they should, for the pure white flowers, the fragrance, the rather useful, slightly running habit, the fresh green foliage that fits in with anything and the modest price; all this makes it very tempting.

Over the years I have often suggested growing

---

*Opposite*

ABOVE Two related plants intermingle contentedly in John Fielding's London garden: 'Bountiful', a large-flowered American selection of *Dicentra formosa*, grows alongside *Corydalis flexuosa*, a recent introduction from China.

BELOW Two easy plants in an attractive combination: Bowles's golden grass, *Milium effusum* 'Aureum', has self-sown itself perfectly alongside that most dependable of dicentras, 'Stuart Boothman'.

hostas and daffodils together so that the developing hosta foliage hides the dying daffodil leaves. But I'm beginning to think that as the tight, clump-forming hostas mature their foliage becomes so dense that the daffodils are deprived of rather too much light, and that their root system becomes too competitive for the daffodils. In large gardens with broad expansive beds everything can be spaced out to balance the planting better, but 'Honeybells' has a more open habit of growth, so that although it spreads more than many at the root it never seems quite so dark under its leaves and the daffodils seem happier.

I have it under an apple tree with 'Buttercup' ivy clothing the trunk and 'Little Gem' daffodils scattered among its roots. Actually I think 'February Silver' might have been a better choice and if one day I split the hosta, as I should soon, perhaps I shall change the daffs.

'Honeybells' is a hybrid of the white-flowered *H. plantaginea*, whose own fragrant evening flowers can be such a feature of the autumn garden. But the parent does demand a hot, sunny spot to flower well, and preferably one which is both hot and not too dry – if you can manage that. Unfortunately it flowers late and can be cut down by the frost; its offspring 'Honeybells' is usually safe. 'Royal Standard' is another hybrid of the same species, but with white flowers, which tends to overdo its inclination to stray from a tight clump to an unnecessary degree; some call it a vigorous spreader.

Another excellent flowering species is *H. ventricosa*, whose rich green glossy leaves set off the deep purple flowers perfectly. This is a toughie and will take full sun if the soil is not too dry, otherwise it is happy in light shade. It is also the perfect hosta to grow from seed for, by a cunning method of reproducing by seed without the necessity for fertilization, a process called apomixis which is common in dandelions and bram-

bles, all the seedlings are identical to the parent.

Discussing a plant of *H. rectifolia* with a friend, a snort sounded from his ten-year-old son, who had recently being studying human biology at school. The true reason for the name is less entertaining – the leaves tend to be rather erect in habit – but again the flowers are the important feature, purple with dark streaks on the insides. I also like 'Tall Boy', which at its best can produce a veritable forest of stems topped with pale mauve flowers, the neat *H. tardiflora*, flowering well into the autumn with purple flowers in dense heads at first but slowly stretching, and of course 'Hadspen Heron', a real star.

'Hadspen Heron' comes from the second generation of the cross between *H. tardiflora* and *H. sieboldiana* 'Elegans' made by Eric Smith. A neat little plant with glaucous, blue-green foliage, the flowering stems are only 12–14in tall, the flowers dark lavender in colour and very striking. This is lovely at the very front where its foliage can overlap a pale, honey-coloured edging stone, although it is rather slow to increase and emerges early from the soil so can be frosted more often than some.

Finally back to *Hosta* species AGS J 424, seed of which was collected in a bog on Hokkaido by the Alpine Garden Society expedition to Japan in 1988. Although in general this expedition proved a little disappointing, I have a few good perennials that came from it and this is one. It has proved to be a neat plant, eventually reaching 24in in height in flower, with delicately pleated, fresh green, elliptical foliage and flowering stems which sneak through the foliage in early June and open to deep purple-blue flowers in July. This year, it's sending up more flower stems at the end of September as the leaves are yellowing.

I cannot tell you its true name; perhaps I should send a piece to the good Mr Schmid. I like it for its modest but effective foliage and the profusion of its dark flowers; and for all I know it might already be in cultivation under its true name.

Many of these flowering hostas are best given space to show themselves off. Not, perhaps, 'grown in distressing round beds upon the lawn or set to fill dank situations at the edge of over-clogged shrubberies', as railed against by Louise Beebe Wilder. But placed so that their self-contained symmetry is not confused by competition from alongside. Alternatively, grow them in pots. In preparation for moving house I dug up my Japanese hosta and set it in a 12in terracotta pot; what a sight.

## Main course

It does its best to get its own back on enthusiasts like Wolfram Georg Schmid, does the hosta. The foliage is, of course, generally reckoned to be the main attraction, and given that the basic shape and colour of the leaves vary relatively little, the hosta manages as much variety as we could possibly expect. After all, there are no lacy or pinnate hostas and the foliage colours do not include bronze, copper or purple shades. So the hosta does a great deal with more or less rounded leaves in the green through yellow to white part of the spectrum, with the addition of its glaucous grey-blues.

But two things serve to confuse the issue. Hostas have a noticeable tendency to sport, either in the garden or while being propagated in the laboratory by tissue culture; so, occasionally, oddities arise and new plants may not be quite what they seem. Add to that the fact that as an individual hosta plant settles and matures in the garden the leaf shape, though rather less the leaf colour, will often change slightly, and between them these two features provide just enough variability to keep the most expert of experts guessing.

But, turning up the heat under the hosta pan to reduce our stock, which foliage hostas are we left with? You would hardly expect me to have seen more than a tiny proportion of the hostas described in *The Genus Hosta*, let alone to have grown them, and every year when I look at the cool, pristine performance of the Chelsea Flower Show display staged by Goldbrook Plants, I think I must have more. But the hostas Sandra Bond displays at Chelsea are, unlike mine in the garden, so remarkably slug-free that when I return home, reality hits me over the head like a mugger's brick and I recoil from over-indulgence.

Perhaps a few words are called for here about Eric Smith, a man who has already appeared in this book in connection with hellebores and bergenias and will be appearing again I suspect. While working at Hilliers Nurseries, Eric Smith took pollen from late flowers on a plant of *Hosta sieboldiana* 'Elegans' and used it to pollinate *H. tardiflora*. The resultant plants were referred to as *H. x tardiana*, although this is not a legitimate name. Eric Smith numbered his selected seedlings in a way which has confused later gardeners coming across his numbers in catalogues. 'Halcyon', perhaps his best known, was numbered TF 1 x 7 but the '1 x 7' does not mean plant 1 crossed with plant 7 as we might suspect. In the full citation T stands for Tardiana, F for Filial (roughly, offspring), 1 means the first generation of those offspring and 7 indicates the seventh selected seedling.

Unfortunately, Eric was a little inconsistent in the application of these numbers; he also gave away and sold seed and seedlings, sometimes under names which they did not really deserve. The British Hosta and Hemerocallis Society has taken upon itself the task of sorting the whole thing out and giving cultivar names to those which have been in circulation only under numbers.

Anyway, Eric Smith was Britain's only hosta breeder of any note and his inspired cross has produced some of our most beautiful hostas, all small blues. I grow three or four and would select, in addition to 'Hadspen Heron', 'Buckshaw Blue' and 'Halcyon'.

In fact 'Buckshaw Blue' came from a slightly different cross, *H. sieboldiana* x *H. tokudama*, and was found as a seedling at Hilliers. The result is a blue-green plant about 18in high with rounded leaves whose rims are often raised upwards slightly; unfortunately the off-white flowers do not open fully, although the foliage is so good it hardly matters, so I cut them off. This hosta has won a number of awards, including the Nancy Minks Award of the American Hosta Society as the best small to medium-sized hosta seen, at Alex Summers' garden in Delaware in particular, on their convention tour in 1987. The simplistic British idea of a Gold Medal suddenly seems foolishly naïve. The larger and more imposing 'Bressingham Blue' is the result of the same cross.

'Halcyon' is one of the most popular hostas in the world, although many of the plants sold under this name have actually been inferior seedlings. A superb plant, whose best blue colour is developed in the shade, its leaves broaden and become heart-shaped at the base as the plant matures. It looks good with Bowles's golden grass seeded around and through it, with blue acaenas straying out from underneath or in a bold border association with 'Emerald 'n' Gold' euonymus.

This is a variety which above all others is best planted and left undivided, for when it is split up it regresses to its more juvenile form and then takes some years to develop its mature elegance.

'Krossa Regal' is also best as a mature clump (I suspect I will be saying this about most of them), and is a most distinctive plant with long, rather upright leaf stems and furrowed leaf blades which strike out at a distinctive angle. And all in a soft glaucous grey-blue. This too is in a pot at present, a 26in terracotta one in which it looks truly majestic. The leafage reaches about 18in high; the flower stems are well over a yard and are topped with lilac flowers. What a specimen. And best left that way rather than crowded too tightly with neighbours.

Perhaps the most popular of all hostas is 'Frances Williams', and this is one which keeps the hosta experts guessing. A sport of *H. sieboldiana* 'Elegans' with an irregular yellow margin to the blue leaf, its development varies slightly in different conditions. It also has a tendency to sport, sometimes to all yellow or back to all blue, and more confusingly to different degrees of yellow margination. Many of these forms have been given names, and not only the all-gold ones, 'Golden Sunburst', but yellow-edged forms which differ only slightly from 'Frances Williams' itself. That these are actually distinct can be an illusion, for the long road to maturity or a move to a new situation in the garden may change them. For all that, 'Frances Williams' is a superb plant which is deservedly popular. And names like 'Eldorado' or 'Chicago Frances Williams' are best ignored.

I pick just two other variegated hostas, and both pose difficulties – actually I'm beginning to think I've already over-indulged so must move on quickly towards the final course. 'Thomas Hogg' is a name used in Britain mainly for *H. undulata* var. *albomarginata* and in the United States mainly for *H. decorata*. As it happens, Mr Hogg the nurseryman has little to do with the British plant and indeed the name is one which is applied with relief to many an unknown white-edged, green-leaved hosta found in British gardens. At least we know that Thomas Hogg was responsible for sending *H. decorata* to the States from Japan in 1884, and from then on it

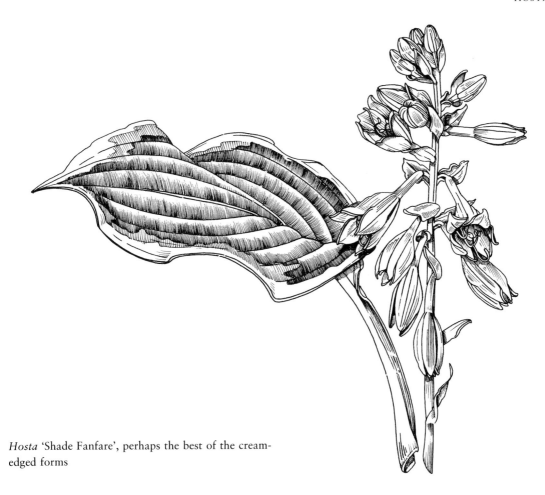

*Hosta* 'Shade Fanfare', perhaps the best of the cream-edged forms

became known as Thomas Hogg's plant.

My plant is *H. undulata* var. *albomarginata* and when I brought it over from Ireland I found tight in the clump a root of *Viola soraria* 'Price-ana' with its dainty white flowers, streaked in blue in the throat. This was a most happy chance, as the violet rushes up to flower early and then the hosta grows up through it. But the pair of them need feeding to match this double appetite.

A more recent introduction is 'Shade Fanfare', and here is the perfect case of what hosta fanatics must contend with and how the rules of horticultural nomenclature must sometimes be bent.

'Flamboyant' is a hosta reckoned to have arisen following X-ray treatment. The leaf is green yet splashed and streaked in cream, yellow and gold; a singularly unpleasant plant. Buy a plant of 'Flamboyant', not that you will find it all that easy, plant it in the garden and in a few years time you will recognize it as 'Shade Fanfare'. For as it matures it loses its central splashing and settles down to a bright green leaf with a slightly creamy yellow edge. And in the transformation of the one individual from a horror to an indispensable plant, it has turned from 'Flamboyant' to 'Shade Fanfare'. So the very same plant is allowed two different cultivar names to cover two distinct phases in its growth.

Perhaps this explains why for some years my two plants of 'Shade Fanfare' looked slightly different. (Actually, their main similarity at present is the number of meals they have provided for slugs.)

Once this plant has settled down to become 'Shade Fanfare' it has such a clean look that if you must have just one pale-edged hosta, this is surely the one. And it is very distinct, with its leaves lapping over each other and their slightly yellowish green blades edged in creamy yellow. Having just inspected mine I confess it does look good with *Carex muskingumiensis* alongside – or it would, had my slug patrol been more vigilant.

Finally, yellow-leaved hostas. Eric Smith had a go at these too, attempting to raise one which kept its colour all summer and flowered well. He used *H. fortunei* var. *albopicta* f. *aurea* (which most of us still call *H. fortunei* 'Aurea'), *H. sieboldiana* var. *elegans* and also 'Kabitan', but never had the success he had with his blues. I've never grown any of these, but 'Gold Haze' and 'Oriana' are said to meet at least partially his aim in that they do keep their leaf colour a little longer than the old favourite *H. fortunei* 'Aurea'.

But please, please, plant none of these hostas with their brilliant yellow spring foliage under magenta-flowered azaleas as they have at Wisley, at least not without warning signs. Driving through the gardens on my way to judge at the Trials Field one spring, I caught sight of this literally breathtaking combination, was momentarily transfixed by the shock – and nearly ran over the Curator.

But take a look at 'Zounds' which, we might say, is a bright yellow version of *H. sieboldiana* var. *elegans*. Although it does fade towards green it remains a superb plant – with an admittedly daft name.

Then there is another American variety with an unusual name, 'Sum and Substance'. The Americans, bless 'em, love giving awards and let me just list those which 'Sum and Substance' has accumulated: Midwest Gold Award 1984, Eunice Fisher Award 1984, American Hosta Society President's Exhibitor Trophy 1987 and the Alex J. Summers Distinguished Merit Hosta 1990. I kid you not, there is also the BBA, the 'Big Bucks' Award, given to the contributor of plants to the auction at the American Hosta Society annual convention whose plants raise the most dosh. 'A plaque,' I am assured, 'is awarded for permanent retention.' Still, I suppose this combination of meritocracy and sales revenue in the hosta world is but the country in microcosm.

But I seem to recall entering into a discussion of 'Sum and Substance'. Apart from its almost statuesque appearance, its yellow or even old gold leaf colour lasts well and this is a hosta which colours best in full sun and can stand the heat too. It has presence, it has colour, it has vigour and it has the one quality everyone looks for, especially those with no special love of hostas in the first place: it is virtually slug-proof.

### Third course

I begin to feel sated; I wouldn't want to over-indulge. And finishing with a hosta which, to slugs and snails at least, is virtually inedible seems appropriate. But I recall my intention to discuss the propagation of most of the plants featured in this book, so thoughts about propagation can round off this meal, as they so often do after a cosy evening in the restaurant.

Commercially, many hostas are propagated by micropropagation in tissue culture laboratories. They respond to this method well, and it is so very much quicker than traditional division. Occasionally sports arise in tissue culture which are often spotted by the lab and retained for trial. Sometimes they slip through, and the trays full of young plants sometimes seen in garden centres are always worth checking.

In the garden, physically dividing the clumps is the only way to increase stocks, but not by using the traditional two-forks method. Instead, examine clumps in early spring when the noses are peeping through the soil and scrape away some soil to reveal the crown in more detail. Then, using a sharp spade on old clumps or a bread knife on younger ones, cut out one or two segments as if taking slices out of a cake. Hostas usually look best as substantial, mature clumps, and by approaching propagation in this way the original plant is left in place, continuing in its elegant maturity. The gaps are filled with used potting compost.

The segments which have been cut out can be replanted as they are, or pulled and cut apart with a knife into smaller pieces which are potted up for a year to help them establish themselves. This disturbance may well return some varieties to a juvenile leaf form uncharacteristic of the mature clump, particularly those pieces split and potted. But eventually, after replanting, or potting on, they will settle and mature to their true character.

Hostas can be raised from seed but the results from all but *H. ventricosa* will be highly variable. Some are actually sterile; those which are not will tend to cross with each other to give highly unpredictable and largely unremarkable results. I raised my AGSJ 424 by sowing the seed in proprietary peat-based seed compost and placing the seed in a propagator at about 70°F. Wild collected seed is sometimes of poor quality and only one seedling germinated; but it did so quickly and your own, home-collected seed will probably come up like cress.

# Euphorbias

There are two types of euphorbias, those with a clear sense of their own identity and those without. In the first group come weeds and the truly herbaceous species; in the second group are those with an identity crisis, pretending to be cacti or unable to decide whether they're herbaceous perennials or shrubs. It is with some relief, for obvious reasons, that the only mention of the pseudo-cacti occurs here; weeds, naturally, are also excluded. And just to be perverse, most of those under review in this section entitled Spring Woodlanders are actually sun-lovers.

## The Mediterraneans

Among the plants in this book at least, the principle sufferer of a confusion of orientation is *Euphorbia characias*, one of the very few plants to appear in both Bean's *Trees and Shrubs Hardy in the British Isles* and in *Perennial Garden Plants* by Graham Thomas. But, as is often the case, those who try their best to overcome fundamental dichotomies in their nature are well loved for the effort and so eulogies to this plant can be found in the writings of Gertrude Jekyll, E.A. Bowles, Margery Fish, Beth Chatto, John Raven, Graham Thomas, Christopher Lloyd and many more.

Technically, I suppose this species should be called a sub-shrub, for there is a persistent woody stem at the base from which arise less permanent branches which die after seeding to be replaced by new shoots from the base.

In the Med it grows in hot, dry places, with the subspecies *characias* tending towards the western end with subsp. *wulfenii* at the eastern end. Back in primary school, the simplest way to remember the difference between the two is to recall that *wulfenii* is without the dark maroon eye which is found in subsp. *characias*. The latter is also generally shorter. As you might expect from a plant which is found from one end of the Med to the other, there is more variation than is indicated by the apparently clear-cut distinction between two subspecies.

I grow quite a few named forms such as 'Lambrook Gold', 'Blue Hills' and 'Humpty Dumpty', plus some seedlings of my own and crosses involving *E. amygdaloides* 'Rubra' to a greater or lesser extent such as *E. x martinii* and 'Perry's Winter Blusher'. These are invaluable and permanent features of my Mediterranean garden, which is made on chalky boulder clay, clunch as we call it in outer Northamptonshire. Here they associate well with other Mediterraneans. The bold, silvered leaves of *Glaucium flavum* gather round the foot of the red-tinted *E. x martinii* ('The green flowers with dark eyes may recommend it,' says Graham Thomas with little conviction), and the flower stems are not so dense that the euphorbia is shaded out. That enthusiastic and long-flowering annual pea *Lathyrus chloranthus*, best from sowing outdoors in the autumn, with its picric green flowers which are so useful in summer posies, sprawls and clambers over 'Blue Hills'. Another year that sometimes unknown or even despised biennial *Adlumia fungosa* drapes its trails over the blue euphorbia foliage.

Hybridization, identification and propagation in this group have provided their mysteries over the years. If, for example, *E. amygdaloides* subsp. *amygdaloides* will cross with *E. characias* to give *E.* x *martinii*, why does *E. characias* not cross with the other subspecies of *E. amygdaloides*, subsp. *robbiae*? And why does *E. amygdaloides* subsp. *robbiae* not cross with var. *purpurea*? Why does *E. characias* cross with one form of *E. amygdaloides* but not others?

Margery Fish was a great enthusiast for these plants, especially subsp. *wulfenii*: 'No weather seems to daunt this handsome plant. Those lovely grey-green spikes remain calm and untouched by the bitterest weather, and if it flowers, it will do so in the winter, when its flowers will be even more welcome than in the summer. Flowers may come and flowers may go, but that handsome mass of glaucous foliage will be there as a foil and a furnishing and a refreshment for many years.' Her favourite was named 'Lambrook Gold', another is 'Lambrook Yellow', and as you might guess there has been some debate about whether they are the same and whether either still exists.

Mrs Fish had a generous nature, but this led her to create problems for those of us who garden a generation or two later. Seedlings from around plants were dug up and given away, sometimes in great quantities, and were grown on under the name of the parent. But of course, asks he rhetorically, will they be true? Certainly many of her hellebores were not – what about the euphorbias?

Mrs Fish showed 'Lambrook Gold' at Chelsea in 1968, the year before her death. She gave away seedlings herself and in 1974 The Plantsmen also listed seedlings; of course we don't know quite how true these were, especially as The Plantsmen grew so many plants in this group on their nursery that there must have been many and varied opportunities for crossing. Indeed they introduced one of their seedlings from 'Lambrook

Gold' as 'Sunstrike' in 1977. Distinguished by being 'taller and with more cylindrical heads, generously tinted with yellow', it seems to have vanished without trace.

'Lambrook Gold' is said to be distinct from others of this sort in its relatively small flower heads and in throwing occasional variegated shoots. Then there's 'Lambrook Yellow' and 'Margery Fish', names used for seedlings of 'Lambrook Gold' which are bound to be variable, although the 'Lambrook Yellow' raised from cuttings by some nurseries is distinct in its bluish foliage and red stems.

And what of 'John Tomlinson', raised at Kew from seed collected by John Tomlinson on a trip to Yugoslavia in 1974 and named after him by Brian Halliwell? I remember being at first startled and then delighted to find this plant used one year in the spring bedding at Kew, with the red campion, *Silene dioica*, I seem to recall. This form is characterized by its early flowering, usually starting in February, its rather lax growth and yellow 'flowers' in almost spherical heads. Of course the colour is provided by bracts which surround the true flowers, and which themselves have often disappeared by the time you come to try to fathom their structure.

Now comes news of 'Jayne's Golden Giant', with greyish green leaves and greeny yellow flowers which individually are said to be larger than those of any other variety.

The Plantsmen also introduced the wonderful 'Blue Hills' in 1974, and they described it as having 'rather mop-headed greenish flower heads and the best blue-green foliage of any here'. It is certainly a penetrating blue, as leaves go, and they are long and attractively wavy. But they lack the delicate down of silver hairs which is such a feature of the short-leaved 'Portuguese Velvet', introduced by John Fielding having been collected in Portugal by a friend.

In their very first catalogue in 1967 The Plants-men also listed *E. characias* 'Compact Form', which reached just 2½ft. Twenty years later, seed from the Hardy Plant Society under this name produced one of the most prolific of my euphorbias. A *wulfenii* type, unlike the dark-eyed 'Humpty Dumpty', it reaches about 3ft at most and never snaps even under heavy snow, as long as the 'Russell Prichard' geranium growing through it is cleared away in autumn. I'm tempted to give it a name and launch it into the world, but there are so many already. Oddly enough, although the seed capsules make quite a noise as they pop on hot June afternoons, not one self-sown seedling of this plant has so far appeared. Seedlings from the others are everywhere.

Perhaps, towards the end of this section on euphorbias with an identity crisis, the time has come to be really nasty, to mock the afflicted, to be rude about the variegated forms; fortunately they are but few. The first I grew, out of curiosity more than devotion, was *E. amygdaloides* 'Variegata'. This was a mistake worth making, once, and I even recorded its sickly pink and cream progress on film – but I will not disturb the restfulness of your repose by including the photograph in this book. Now some are saying that this form should be called 'Frailty'; charity has its limits.

How the same colour on the same plant can be called both silver and gold defeats me, especially when it's cream. At first named 'Benger's Silver' after Judy Benger, who found this variegated seedling of *E. characias* in her Devon garden, by the time it had been released on the market its nomenclature had progressed through 'Silver Sunbeam', the preposterous 'Honiton Lace', and now rests as 'Burrow Silver' after her home, Burrow Farm. I last saw it at a plant sale as 'Barrow Gold'; perhaps this was someone trying

to sneak a way round the Plant Breeders' Rights legislation.

This plant is certainly bright, with broad leaf margins varying in the mature leaves from cream through to primrose, and is a brighter yellow in the shoot tips. Occasional pure yellow shoots appear but cannot be grown on separately; occasional green ones come too, and must be cut out. Seedlings are almost always green, the ones surrounding plants in the Bressingham stock beds exclusively so.

To be honest, having been out to inspect my own plant I find myself less inclined to be rude than I thought I would be. I quite like the pink stems behind the pale leaf margins but I'd like it more if it spent nine months of the year green and developed a seasonal variegation. Cuttings seem to root easily, but growth is slow. This would look wonderful wreathed in *Lathyrus chloranthus* but the weight would probably break the rather fragile branches; *Nicotiana langsdorfii* growing through would be fine. A winter later, translating these ideas into reality will have to wait. The plant has succumbed to an average season's cold and wet.

'Burrow Silver' is a form of *E. characias* subsp. *characias* which arose as a seedling of a less well-defined variegated plant. The more recent 'Emmer Green' is a form of subsp. *wulfenii*, and although the pot-grown plants which I've seen are smaller and neater than 'Burrow Silver', in gardens it appears more vigorous and its cream edge more clearly defined.

To be honest, the foliage of the nice spurge, sorry, the Nice spurge, *E. nicaeënsis*, is better than all these; its crowded leaves are a lovely pale greyish blue and there are limy flowers all summer.

So many, yes. Roger Turner's diligent researches for his recently published monograph have yielded more than two dozen cultivars which are described in his book, and it looks as if more good ones are on the way, especially from Gary Dunlop in Northern Ireland.

### Mediterraneans from seed and cuttings
As has been said, growing these euphorbias from seed will not give true plants; the very fact that improved forms turn up as self-sown seedlings indicates that offspring of any particular plant will vary and may not be identical to the parent. Unfortunately there is no escape from the fact that while many will be similar and one in a million might be better, the majority will be worse.

Fortunately they are easy to raise from cuttings. Young shoots 2–3in long, though not the tips from mature shoots, can be taken at almost any time of year, though preferably in spring and early summer. They should be rooted in an exceptionally gritty compost – four parts sharp sand to one part peat is good, and I've even used pure vermiculite. Cuttings should not be placed in a propagator but in an airy place in the greenhouse out of the sun – under the bench is ideal if it's not too shady.

### Hardy and truly herbaceous
The only euphorbias without an identity problem are these – although some pose problems of identification for us instead. Even those which are not truly herbaceous look as if they are; they fall into two groups.

As a companion for spring bulbs and an early feature in perennial borders, E. polychroma has long been grown. It is not truly herbaceous, dying back to a short, wiry, woody stock each winter. This is an intriguing plant, whose picric yellow flowers open as the buds push through in March and continue to open as the stem stretches so that new buds are still opening as the stems reach 18in in height.

I grow 'Purpurea' with warm, smoky purple bracts around the yellow flowers, and this is better with blue scillas than the ordinary yellow form. By contrast, 'Sonnengold' is an exceptionally bright yellow but rather slower-growing form, looking very sophisticated surging through the paler 'Allgold' lemon balm – until the 'Allgold' frazzles at the edges in the late spring sun. Cut back the 'Allgold' to make a good foil for the dark, red-tinted, blue-green leaves of the euphorbia.

One plant which swept round the country quicker than almost any other I can recall was Euphorbia dulcis 'Chameleon'. Found in the wild in the Dordogne, the leaves are said to be green in the sun and reddish purple in the shade; sometimes opposite sides of the same plant show contrasting colouring. Some plants seem to remain reddish purple in the sun until the whole plant turns brighter red in autumn.

Originally thought to be a hybrid between E. dulcis and E. villosa 'Purpurea', it now seems that this is a tall, dark-leaved form of E. dulcis; it does not come true from seed. If nurseries raise plants such as this from seed and refuse, or are too greedy or are too ignorant or are simply too lazy to select plants for sale according to the characteristics of the true variety, they are simply swindling their customers. 'Chameleon' should either be rigorously rogued before being sold or propagated vegetatively.

Having said all that, at its best this is a wonderful plant and ideal for enlivening an autumn border in spring before contributing its own fiery finale.

The most completely herbaceous group of all contains mainly Asiatic species. I am already weary of saying that this group or that group is rather muddled, but I must say it again here. I realize that I cannot always resolve these little local difficulties to my, or indeed your, satisfaction, but at least you're warned.

The common features of this group, apart from being herbaceous, are their bold presence, their elegant upright stems, their long, narrow, veined leaves and their steadily increasing, sometimes running, habit.

*Euphorbia longifolia* is one of the most familiar names among this group of spurges but, unfortunately, a number of different plants have been given this name over the years – so to avoid confusion, and abandoning the habits of a lifetime, the botanists have decided that none shall have it. Probably the only plant which should have been called *E. longifolia* all along is that which reaches a rather top-heavy $3\frac{1}{2}$ft, with reddish young shoots in spring developing into red stems with long, narrow leaves each with a neat white midrib. The flowers, in June and July, are a slightly greeny yellow and its movements are restrained.

This is now known as *E. donii* and is an excellent plant. Blue-flowered carpeters like *Anemone blanda* 'Atrocaerulea' (now 'Ingramii') look superb with the red shoots surging through them early in the year. Later, bold hostas in blue like 'Bressingham Blue' or yellowish green like 'Zounds' or 'Piedmont Gold' make quite a spectacle. A form called 'Amjilassa', after the village in Nepal near where seed was collected by Ron MacBeth in 1989, is taller at a spectacular 5ft and with larger flowers.

*Euphorbia cornigera*, with similar flowers, also fits in here and has been distributed as *E. longifolia* and also by Blooms of Bressingham as *E. wallichii*. But this is a distinct, usefully later-flowering species reaching a little over 2ft, with those reddish stems and leaves with their white midribs which create the impression that these species are all the same. *Euphorbia wallichii* itself is usually a little shorter and its flowers are past

their best by the time *E. cornigera* gets into its stride.

Quickly now, let's finish these off. *Euphorbia sikkimensis* is the one with the pink colouring at the base of its white midrib, its secondary leaf veins picked out in white and whose roots have a tendency to trek. It flowers late too, well into August. Like *E. donii* the young shoots are bright red but have a way of emerging very early, then waiting until the weather warms up before stretching. The result is that *Scilla siberica* 'Spring Beauty' is a better companion than the more overwhelming anemones.

Then there's *E. schillingii*, collected by Tony Schilling in Nepal in 1975 and originally thought to be a form of *E. sikkimensis*. In fact it differs in its large, grey-green foliage with a white midrib and its even later flowering season, which stretches into September.

Finally, perhaps the best known of all these, *E. griffithii*. Rather athletic, in light soils especially, the best forms are stunning. Flowering from April onwards, new flowering shoots branch out from below the first head to overtop it as it fades, increasing the height and lengthening the flowering season.

Never order *E. griffithii* 'Fireglow' by post. Because it has been raised from seed over the years, you can never be quite sure what you will get. Instead go for 'Dixter', selected by Christopher Lloyd from seedlings at Washfield Nursery, for its reddish foliage and flowers in soft burnt orange. It is shorter and less likely to wander than others. Look too for 'Fern Cottage', with its slightly fierier flowers and fiery autumn colour to follow.

These are both good floating in pools of green ferns or against green-leaved shrubs, and a neighbour of mine has a spreading clump attractively intermingled with *E. amygdaloides* var. *robbiae*. For more dramatic tastes, a background of

*Euphorbia schillingii*, discovered in Nepal in 1975

yellow-leaved physocarpus or philadelphus and a surrounding of doronicums might suit.

Seedlings of some of these may spring up in the garden, should the plants be happy, but beware that they are not inferior to the parent. All those Asiatic species can be divided in early spring; the soil falls off the slightly alarming white rhizomes but they soon settle. Cuttings are also a possibility, but usually an unnecessary one.

### Closer to home

There are more, many more, euphorbias which I could mention and I'm sure that Mediterranean species like *E. ceratocarpa* will become increasingly widely grown. But perhaps these thoughts should be rounded off with mention of *E. amygdaloides*, the British native wood spurge.

Only a few months ago I was wandering about in the New Forest looking for *Pulmonaria longifolia* when I glimpsed a golden-tinted green haze through the birches and it proved to be a vast colony of *E. amygdaloides*. It seemed to grow where the canopy was perhaps thinner than elsewhere and was an entirely evenly spaced and uniform stand, with just a few scattered outliers, as if the whole area had been taken over by a single relentlessly spreading plant. In a ditch near by a good red form grew, which I think we must assign to var. *purpurea* and which looks to be as dark as my garden form.

The two plants which featured on that New Forest exploration suggest themselves as good garden companions, the red-leaved form of the euphorbia and *Pulmonaria longifolia*, perhaps in the prettily spotted, pale-blue-flowered form 'Roy Davidson'. But this does depend on having a good dark form of the euphorbia, and one of those interesting tasks assigned for 'when I have time' was to select a richly coloured, dark-leaved form of this plant which is also resistant to mildew. It seems that I may have been overtaken in this intention by Gary Dunlop, but I wonder if in Northern Ireland powdery mildew is so much of a problem as in eastern England where the summers are so dry.

Arguments persist as to the right and proper status of var. *robbiae*, some favouring its elevation to subspecies or even species status. There was a time when I thought this a very boring plant, but I'm coming to like it. Deep emerald green foliage topped with heads of flowers that Mrs Fish described as 'love-bird-green'. Like others in this group, the stems with their rich foliage grow all season and then flower the next. *Flora Europaea* mentions subsp. *semiperfoliata* from Corsica and Sardinia which sounds intriguing; the stems flower in their first year. Next time someone's in those parts…

There are hybrids among these plants, in particular crosses between *E. characias* and *E. amygdaloides* which are known as *E. x martinii*. There are two wild forms and two garden forms of this cross in cultivation, but only one is generally available and this is correctly known, it seems, as var. *pseudocharacias*. Lord Talbot de Malahide once made a deliberate cross between *E. amygdaloides* var. *purpurea* and *E. characias* subsp. *characias* and this inspired idea resulted in a plant named, what else, 'Malahidensis'. It seems to have been lost since his death, and needs trying again; how do you hand-pollinate a euphorbia?

*Euphorbia* x *martinii* is one of my favourites; my plant at least is stocky and compact, keeps its deep green leaves right down to the base of its stems, and its dark-eyed green, rather than yellowy green, flowers are very striking. Like all these sub-shrubby types it suffers from aphids, but it looks wonderful with the rosettes and then the flowering stems of *Galactites tomentosa* or *Glaucium flavum* in front. In my garden self-sown seedlings of the latter have now displaced those of the former.

# Pulmonarias

Pulmonarias cause problems for gardeners. Fortunately the least of these difficulties is actually persuading them to grow, for they could hardly be more amenable. They clump up well in shade and in sun too if the soil is not too dry; they are easy to split and they self-sow happily around the garden. No, confusion arises simply because so many are so very similar.

### You thought hellebores were confusing...
Hellebores, with the large number of very similar named and unnamed forms and their tendency to self-sow seedlings which are completely different into their own clumps, cause endless confusion. Surely pulmonarias could not be worse? They could and they are, for, like primroses, pulmonarias are equipped with pin and thrum flowers with the specific aim of encouraging cross-pollination. So, you might conclude, most self-sown seedlings will be crosses with other plants. You might also reasonably conclude that self-pollinated flowers will, like primroses, yield but a small quantity of poor seed. Not so fast. Some pulmonarias are strongly self-fertile and yield plenty of seed. Some are entirely sterile to both their own and other pollen and produce no seed at all. Some have infertile pollen so all their self-sown seedlings are hybrids.

This leads to anything from the complete absence of seedlings around a particular plant to a mass of seedlings whose parentage will never be entirely certain. So, to sum up, some plants are male sterile and some are female sterile, some are fully fertile and some are completely sterile. There are plants with pin-eyed flowers and plants

with thrum-eyed flowers; there are even plants which seem absolutely identical except for that difference in flower structure.

The most significant result of all this is that until Vanessa Cook has completed her study of her National Collection, we must view any plant we buy from a nursery with a certain amount of circumspection. We must also learn to dead-head our own plants before their seed is shed to eliminate the possibility of seedlings germinating in and around our clumps. So many seedlings turn up in gardens that we must resist the temptation to name anything without first ensuring that it is not only a good plant but distinct from, and also better than, the many others already around – in either leaf or flower. Got all that?

### The simple life
There was a time when we all grew what on both sides of the Atlantic we called simply lungwort, *Pulmonaria officinalis*. The flowers opening in March or April are pinkish at first, turning to pinkish blue, and these are followed by rough, spotted leaves. This has long been a cottage favourite, from the days when Robert Turner asserted that the appearance of a plant indicated its medicinal use and the spotted leaves of the pulmonaria, boiled in beer, were deemed the medicine for lung complaints.

This Doctrine of Signatures would indeed be a most advantageous arrangement. I could hardly resist the instant application of a Victoria plum to the nose of a friend visiting today and complaining of hay fever. But no, God in his wisdom has been more obscure in his distribution of

69

valuable curatives through the plant world; I am as yet unclear as to how the signature of its virtue is imprinted upon the willow whose bark relieves headaches by the action of the aspirin it contains.

Anyway, this pulmonaria was introduced to Britain from mainland Europe, although the rather similar *P. obscura* is native and clings on in a few Suffolk woods. The altogether more attractive *P. longifolia*, with longer and more strikingly spotted foliage (at least in garden forms) and purer blue flowers, is a little more widespread in Dorset, Hampshire and the Isle of Wight. But all are increasingly rare in the wild in Britain.

In the New Forest *P. longifolia* is found in a number of locations, though sparsely and decreasingly in the two I've visited. In both sites the plants were under threat, in one from increasingly dense scrub so that in many areas the plants had been smothered by brambles and in the remaining relatively open areas they were small and scattered. In the other, logging had removed cover then provided it anew in the

form of stacked logs while between times the machinery had churned up the soil.

In neither location were the leaves more than thinly and lightly spotted, indeed the foliage of some plants was so little marked as to be closer to *P. angustifolia*. But populations clearly vary, and did in Farrer's time otherwise *Pulmonaria longifolia* could hardly have been his favourite: 'Pulmonaria will not easily find a lovelier representative than the narrow-leaved brilliant Spotted-dog of the Dorsetshire woods, with 6- or 8-inch stems, and its hanging lovely bugles of rich clear blue in April – so much more modest in leaf, well-bred in growth, and brilliant in flower than the towzled and morbid-looking heaps of leprous leafage made by the common Lungwort of gardens, with leafy stems and indecisive heads of dim pinky-blue flowers that look as if they are going bad.' He continues in the same indefatigable vein and I wonder how ecstatic he would have been over the flowers of *P. longifolia* 'Bertram Anderson' or how scathing of the foliage of *P. mollis*.

The foliage of *Pulmonaria longifolia* 'Bertram Anderson' (*top*), *P. saccharata* 'Argentea' (*middle*) and 'Lewis Palmer' (*underneath*)

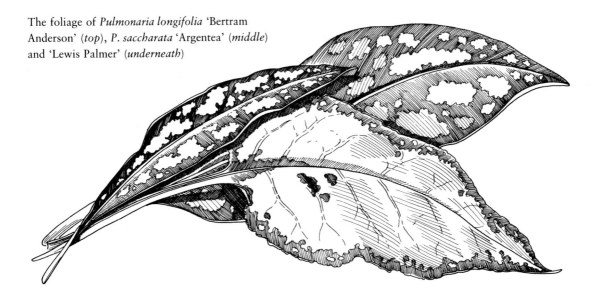

*Pulmonaria saccharata*, with its more strikingly spotted leaves, has been the other long-term occupant of our gardens, but in the 1930s nurseries specializing in perennials sometimes omitted pulmonarias from their catalogues altogether in favour of more flamboyant plants. Now, hundreds are grown. A few are hybrids but many are selections chosen for the purity of their flower colour, the absence of the colour change from pink to blue, for the size of flower and for their foliage. And as they became increasingly fashionable, any seedling which turned up in a garden and which was different from the plants already growing there may have been given a name – in spite of the fact that similar named plants may already have been established in other gardens.

There was a time when I considered collecting pulmonarias with a view to clarifying the distinctions between apparently identical varieties and selecting favourites. 'I have a great liking for them all, and have collected together all I have met with that show any variation, and the working out of the assemblage is one of the jobs I have in store for that day when I shall have some spare time. Perhaps a broken leg might fix the date, but at present it does not appear on my list of engagements.' So wrote Mr Bowles before the First World War and I don't think he ever had a leg-mending period in which to sort them out. I have to say that without the dubious benefit of a fracture I have so far only managed to sort out the four common forms of *P. angustifolia*! Even then, I now suspect that I may not have started with the correct plants in the first place! Vanessa Cook with her National Collection and the Pulmonaria Group of the Hardy Plant Society are doing far better, and I will confine myself to appreciation and encouragement of them in their work and to making good use of their early conclusions.

### Spotted dog, soldiers and sailors, thunder and lightning, Mary's milkdrops

Any plant used in medicine or grown for centuries in cottage gardens acquires common names, although they sometimes seem rather fanciful or obtuse. Farrer's spotted dog refers specifically to *P. longifolia*, although it has also been used for *Orchis mascula*, with its purple spots on fresh green leaves.

Most names refer to *P. officinalis* and fall into three groups. There is a group of names which concern the newly opening pink flowers and the more mature blue flowers seen together on the same plant. These are aligned with the traditional notion of 'pink for a girl, blue for a boy' to give names like Adam and Eve, Josephs and Marys, soldier and his wife. Perhaps thunder and lightning also fits here. The other main group of names refers to the spotted foliage, for it was once widely supposed that the Virgin's tears, or her milk, had fallen on the leaves and spotted them. Hence names like the straightforward Virgin Mary's milkdrops, Virgin Mary's tears or simply Virgin Mary and also spotted Virgin and spotted Mary.

Some names refer to the vague resemblance of the flowers to those of the cowslip, *Primula veris*, at least inasmuch as the flowers come in spring and hang down at the top of a short stem. So we have blue cowslip, Virgin Mary's cowslip and Jerusalem cowslip, Jerusalem being a word sometimes applied to unusual variants of more familiar plants.

### The whites, the blues, the pinks and the reds – and the purples

Basically, the flowers of most pulmonarias except *P. rubra* open pink and turn blue. But the exact shades of pink and blue and the extent to which they remain one colour or the other varies enormously. It is perhaps also worth saying that pin-eyed flowers tend to be larger than thrum-eyed flowers. Let us make a survey, from white to the deepest blue and red, and see which varieties stand out; not, at this stage, taking too much account of the species to which they belong.

'Sissinghurst White' is superb, with pure white flowers held over good dark leaves. This is an *officinalis*, though sometimes listed as a *saccharata*, but I also like the white form of *P. rubra*, known as 'Albocorollata', as the flower looks so well against the paler fresher foliage. I recently saw a young plant of the white form of *P. longifolia* which looked very tempting. A white form of what was reckoned to be *P. angustifolia* was said to grow wild near Newport on the Isle of Wight; it sounds an essential plant.

The palest blues I like the best are *longifolias* or hybrids thereof, in particular 'Merlin' and the fractionally darker 'Mrs Kittle', and although these do have rather small flowers it would not be patronizing to call them dainty. Next comes what I grow as 'Bowles' Blue' plus 'Roy Davidson', both the same shade but the former an *officinalis* and the latter a *longifolia*. 'Bowles' Blue' has rather small flowers and is also slightly paler, with perhaps a touch of lilac, 'Roy Davidson' is a good pale shade but again the flowers are small and seem to stay half closed. The latter makes a wonderful plant with the flowers set against such heavily spotted little leaves. 'Blue Mist' is the most delightful, very pale shade, and while some plants around are the same as 'Bowles' Blue', others are not.

Continuing deeper in colour there comes a real star, 'Frühlingshimmel' from Heinz Klose – which I had better say here and now is the same as 'Blauhimmel'. The large open flowers are a shade which it is difficult to describe. It's a strong blue though not a dark blue, rich but not deep, a sort of heavy Cambridge blue perhaps, but a

wonderful plant nevertheless. If only its summer leaves were better – the spotting is feeble.

Darker still and we come to those dark blues, the best of which really sparkle. 'Blue Ensign', a very large-flowered form or hybrid of *P. angustifolia* found at Wisley, stands out in every way; the flowers are so large and so rich yet they shine so brightly, and the plant is taller than other *angustifolia* varieties. 'Bertram Anderson' has flowers in a similar shade but this is a plant I find surprisingly miffy; it just seems to resent disturbance in a way others do not. But the most familiar is 'Lewis Palmer', which we must say once and for all is the same as 'Highdown'. And this is superb. Must I say more? No. Except that there is a plant going round as 'Boughton' and as 'Valerie Finnis' which is probably the same.

Then there are the pinks and reds, of which there are relatively few, although more are appearing. I don't know of a pale blushed pink; 'Dora Bielefeld' is the best of the mid pink shades. The flowers are not huge, but in my garden at least they come in huge quantities; unfortunately the foliage is exceptionally poor. Then we have the various red forms of *P. rubra*. What can I say? Suffice it to recommend that you only need one of 'Bowles' Red', 'Redstart' and 'Barfield Ruby', which are all a rather similar reddish brick shade. And just the one too of 'Ann' and 'Barfield Pink' – which are no more than pin and thrum forms of the same red-flowered form with its pale streaks in the flower.

Some of those with pink in their names, like 'Beth's Pink', are really quite dark, and among this group I suggest 'Diana Chappell', once distributed as 'Peter Chappell's pink seedling', and for no better reason than that I have grown it and like it, both flowers and brightly spotted foliage, while the very similar 'Beth's Pink' and 'Leopard' I have only seen. Then, getting darker again, we approach the purplish shades which tend to be hybrids between *P. longifolia* and *P. rubra*. 'Mournful Purple' and 'Patrick Bates' belong here but I like 'Cleeton Red', a deep and dusky shade.

### Morbid beauties

When I was about three we had a small, white, blotchy terrier which my father christened Lucky, having rescued it from an early grave at the dogs' home. Unfortunately for Lucky, he celebrated his reprieve with such determined enthusiasm for the local ladies that his ignominious departure was merely postponed, and, having demolished fences and gates in frenzied pursuit, he met his end at the vet's. Unfortunately, whenever I hear the name 'spotted dog', I think not of pulmonarias but of unlucky Lucky.

Fortunately, the plants themselves spark no anguished recollection of a randy terrier but nevertheless the name leaves me ambivalent: spotted dog. Farrer seems equivocal too, though in rather a different way. On surveying a population of what he took to be *P. officinalis* near his hotel at the foot of the Boréon valley in France: 'On their hispid dark ground of greyish green some of them are spotted, others splashed or flaked or patterned with a livid white which looks as if their surface had been blasted with vitriol; in one form, even, almost all the leaf was white, with only a narrow rim of normal colouring round the edge. And these monstrosities have a beauty and a wonder, though the beauty be morbid, and the wonder merely an agapeness that such apparently artificial developments should here be thriving in the unsophisticated wilds.'

This enormous variety in leaf spotting and leaf shape is one of the attractions of the pulmonaria. With two distinct appeals, flowers and foliage, and both in such variety, it's hardly surprising that there are so many varieties – and so many

favourites. But concentrating on foliage now, let me survey a selection of good leaf forms from the entirely unspotted to the entirely white.

*Pulmonaria mollis* – I grow 'Royal Blue' – is like rough and rasping spinach. Very vigorous and strong with surging new shoots standing stoutly upright in the middle of the clump, the puckered leaves are completely unspotted and have dark, proud veins which turn rusty in colour when rubbed by other leaves in the wind. A very bold and assertive plant for the middle distance.

As far as their foliage is concerned, most of the forms of *P. rubra* are rather similar, bright green in colour darkening slightly with age, soft but with a slightly rough surface and, with 'Redstart' at least, unspotted. 'David Ward' is noticeably different in that each plain green leaf is edged in white. Found by Beth Chatto's propagator as a sport in the garden, this form is best out of full sun otherwise the leaf edge tends to crisp and scorch.

'Reginald Kaye' is a very striking form of *P. saccharata* with large distinct spots over most of the leaf except for the edge, where smaller spots are gathered into a speckly rim. 'Merlin', on the other hand, is probably a hybrid of *P. longifolia* and is very prettily spotted but with a distinct green stripe along the midrib. 'Mrs Kittle' too has a green central stripe and long broad leaves, broader than 'Bertram Anderson' or 'Dordogne', with big shimmery silver blotches towards the centre of each half and a distinct green edge with smaller spots. 'Bertram Anderson' is a *longifolia* with fewer, larger spots than 'Merlin' and no green stripe, but seems especially prone to mildew.

As the silvering becomes more striking we come to the superb 'Mary Mottram', in which the spots have coalesced into a silver leaf with just a slight hint of green but breaking into a narrow green edge speckled with silver spots.

'White Leaf' seems to be very similar, if not identical. The *longifolia* version of this type is 'Dordogne'. Then we come to 'Tim's Silver', which is almost wholly silver, though a slightly creamy silvery shade, with a few irregular green specks and blotches and the finest of green picotee rims. And finally to *P. saccharata* 'Argentea', rather a variable plant but pure silver at its best, and then to 'Margery Fish', now reckoned to be an entirely silver-leaved form of *P. vallarsae*.

### Taking control

As I have explained at the start of this chapter, pulmonarias have a habit of propagating themselves entirely according to their own unpredictable inclinations, with but little regard for what the gardener expects or indeed desires. So to be sure of our expectations being fulfilled, intervention is necessary. Translated into plain horticultural language this means dead-heading promptly as each flower head fades and propagating vegetatively.

Division is not difficult. As with many spring-flowering plants, splitting in autumn is successful, but it is perhaps wise to break plants into larger pieces with some old root to provide reserves. When splitting in spring after flowering, plants can be broken into smaller pieces with young vigorous roots retained at the expense of tougher rootstock. Foliage should be cut back and the splits never allowed to dry out.

You may well find that if you move a plant or a clump to a different part of the garden, by the following year it seems as if you never moved it at all; this reveals the useful fact that some pulmonarias can be grown from root cuttings. The *longifolias* seem especially easy by this method, together with those hybrids with *longifolia* blood.

# Primroses

Christmas is looming and I've already stepped out to look at the primroses. My heavy soil is too soggy to step on so I stretch from the path to part their leaves; the wind catches at the small of my back. But there they are, the neat, pointed buds clustered at slug level like a rocket battery, waiting for the warmth to launch them into the spring. Above, an early red hellebore has been flowering since the end of November and alongside are the first snowdrop noses, testing the air; more potent missiles.

The primrose. 'Could anything more delicious be asked for or imagined?' wondered the American Louise Beebe Wilder, an inspiration on a par with Jekyll but almost unknown in the UK. When she moved house in the early 1920s she chose particularly a site where primroses would thrive, much to the mystification of the estate agent. In her new garden one of the first features she created was the Primrose Path beneath cherries and crab apples, which she called a natural mosaic of primroses with scillas, grape hyacinths, snowdrops, dwarf daffodils and violets.

This mosaic is not at all natural. If you walk the Devon lanes, or even lazily drive along them, you will see the wild primroses speckling the high banks. In some lanes it's clear that they are more dense and more luxuriant on the north-facing side. Further north, in Scotland, they grow on open hillsides, again usually north-facing.

But rarely do they grow naturally in a mosaic of other spring flowers; the British countryside does not provide sufficient spring species which enjoy a similar habitat. So in the wild it is a simple and solitary spring optimism which captures our devotion, though in gardens they are content as part of the mosaic. Like trusting children they open their faces to us: fragile, tougher than perhaps we suspect, but completely undone by lack of care. Nurture these wild treasures, but not by constant interference, overprotectiveness and meddling. Rather they need the right environment, and they need it to be maintained, they need certain definite attentions and they need help when they're in trouble. They may survive meddling but they will never flourish.

### A nervous and intractable temperament

This is where I feel guilty, but I was relieved to discover that I am not alone. In an appendix to Sacheverell Sitwell's wonderful classic *Old Fashioned Flowers*, a certain Eda D. Hume contributes a note on the cultivation of double primroses. She begins: 'I feel diffident about writing "How to grow and keep the Old Double Primroses" because to my great regret I have lost all mine'! What perfect qualifications for an expert on primroses – she grew them so badly that they all died. Having outlined the various causes of their sad demise, she concludes: 'I would never attempt them again.' By no means have I lost all mine but I have lost many, and many of the others do little but languish. I will copy Eda D. Hume and allow you to learn what you will from my own misfortunes – not the least of which is to live in a part of the country in which primroses seem naturally discontented.

There was a time when choice primroses, including the doubles, were grown as an under-carpet for tulips in spring bedding schemes, and while this would present an extraordinary spectacle today, a more magical if less immediately stunning scheme in these days of natural gardening is to imitate Mrs Wilder's mosaic.

Opinions differ as to the conditions garden primroses require. Roy Genders, who grew an enormous collection, praises 'their ability to flower to perfection under all conditions'; every so often in Roy Genders' books you come across some outlandish assertion and this is one such. Margery Fish tells us: 'I take great trouble over my primroses, but I often think they would do just as well if I did not fuss so much.' She may be right, but she gardened in cool Somerset. Sitwell insisted: 'It is, perhaps, the melancholy truth that if sufficiently cared for they will always survive. There could be no greater mistake than to imagine they are capable of looking after themselves.'

So where does all that leave us? I will tell you. It leaves us knowing that primroses and the choicer polyanthus require a moisture-retentive soil and an absence of scorching sun and drought. And if the natural climate fails to provide such of these requirements that the primrose is satisfied, then the gardener must. E. A. Bowles gardened on what is now the northern outskirts of London, rainfall about 25in: 'They [primroses], poor dears, are not very happy here except in wet seasons... Short of digging a ditch for them I fear I must not expect to see them thrive here.'

Now back to my own experiences. When I moved to my present garden, rainfall about 25in, the only mature trees to provide shade were the remnants of an old orchard, plums and apples mainly, and primroses and other spring woodlanders demand shade in my part of the country. As there is little point even getting out of bed on spring mornings unless there are primroses and anemones to admire, somewhere had to be found for them.

I made two beds under apple trees; one bed was against a west-facing wall with a tree at one end, a second was directly under another tree. On the wall behind the first bed I planted the pink-and-white-flowered *Chaenomeles* 'Moerloosei', which by its second year was producing new growth 4ft long. In the northern corner was set *Viburnum tinus* 'Gwenllian', with its happy tendency to flower from autumn to spring in company with its blue berries. The apple tree overhung the south end and *Daphne laureola* 'Margaret Mathew' went in the centre.

Among spotted pink and spotted white hellebores from Washfield Nursery I planted a range of the modern double primroses, 'Miss Indigo', 'Marianne Davey', 'Ken Dearman', 'Sunshine Susie' and so on, all selections from the Barnhaven strain, plus some oldies like 'Old Double Sulphur' and 'Lilacina Plena'. The most successful was 'Sue Jervis', in an unusual pinky beige shade which, I must say, has its detractors. This is always said to be one of the more difficult to grow of modern types (can it really have been found growing wild in a wood in Shropshire?), with a particular liking for heavy soil. So I made a special pocket for it, forking in liberal quantities of sterilized loam bought to make potting compost. This is the only primrose in the bed to thrive wholeheartedly.

In three years, from a plant in a 3in pot, it has made a clump 15in across and is stunning in front of *Euphorbia amygdaloides* 'Rubra'. After its next flowering I will split it; we will come back to this. While some of the others cling on, most, in particular the more modern varieties, have gone. There are two particular causes, drought and vine weevil – the two reasons that primrose-growing in Britain has been in a state of crisis.

Those modern doubles came from a well-respected alpine nursery and as they declined and were examined for ailments, vine weevil grubs were found munching through the roots of each one. A friend recently pointed out to a nurseryman the notches in the leaves of the prims he was selling on a market stall. 'Oh, it's only vine weevil,' he said confidently, 'you don't need to worry about that.' Such complacency makes me fume. This is one of the most terrifying pests in the garden, its gluttony never satisfied by destroying our precious cyclamen, auriculas, primroses, heucheras, tiarellas, begonias and even pelargoniums. The grubs eat tubers, roots, stems from the inside so that until it is too late and their sickly state alerts us, or a gale simply blows the plant away leaving its severed roots in the soil, we are contentedly unaware. Or we were. Now I worry all the time that the little beasts might be scoffing my 'Lilacina Plena' in the same way they did my 'Sunshine Susie', 'Barrowby Gem' and 'Dorothy'.

This problem has grown significantly in recent years as a number of effective but singularly poisonous chemicals have, quite rightly, been taken off the market. Parasitic nematodes have generally proved useless, but something new and effective is now on the market for nurseries, though not yet for home gardeners, and although it's a strong and long-lasting organo-phosphorus insecticide it does work; so we may soon be able to sleep a little more easily at night. At present, the answer can only be to remove all the soil from newly bought plants, wash the roots in systemic insecticide and repot them into your own, fresh, pest-free compost. Why 'Sue Jervis' has so far proved resistant is perhaps due to the unusually heavy soil in which it grows, and perhaps because outside in the garden vine weevil grubs take some time to develop into adults and spread.

The other bed under the apple tree also has a daphne, *D. albowiana*, as well as *Sarcococca confusa*, but good leafage is provided more by some beautifully marbled *Cyclamen hederifolium* with arrowhead leaves, *Bergenia* 'Morgenröte' and three hellebores including *H.* x *ericsmithii*. There are snowdrops including the wonderful 'Merlin', with its entirely green inner petals, and the broad-leaved *G. ikariae* var. *latifolius*. And there are primroses and polyanthus: Carol Klein's 'Rhubarb and Custard', the true 'Prince Silverwings' and 'Guinevere'. They are in heavier soil than those in the other bed, which is made of cottage garden soil with nearly two hundred years of mucking, but they have been mulched every winter in the approved manner; these too are declining, even 'Guinevere'. The reason is drought, and in other parts of the garden the same is true.

Drought encourages the decline of these choice primroses and polyanthus in two ways. It provides ideal conditions for the deadly predations of the red spider mite, which destroys the foliage and thwarts the plants' efforts to make fresh roots and new crowns after flowering. And of course at the same time the very lack of moisture itself restricts growth severely. This is most noticeable when plants are divided. The theory is that the plants are lifted, split and replanted in improved soil immediately after flowering. But dry summers then attack them at their most vulnerable and they can disappear very quickly.

The gardener can help. In areas where prims are naturally difficult, grow them on a north-facing slope or ditchbank, under a high canopy of branches or among the perennials in the herbaceous border where they are open and exposed in spring but shaded by taller plants all summer. Roy Genders planted tall annuals among his prims for summer shade. Water them if the water company will let you, and mulch every autumn so that when growth starts the new roots have something good to grow into.

Regular splitting and replanting is usually recommended and this is good advice, if only because it enables the roots to be searched for grubs and treated before replanting. In some areas, with some varieties, dividing and replanting every year is necessary, but this is a very fussy and time-consuming business which deprives us of ever seeing a substantial clump of anything. And a practice which brings us back to spring bedding, an inconvenient resting place for a book on hardy perennials.

Some primrose varieties are undoubtedly tougher than others. 'Alba Plena', 'Old Double Sulphur', 'Guinevere' and 'Wanda' seem especially tolerant of less than ideal conditions

and while I have lost 'Gareth', John Fielding's superb hybrid between 'Guinevere' and a red Cowichan, many gardeners find it unusually resilient.

Or, of course you could raise them from seed. Now that Barnhaven primroses are re-established in Brittany, seed in a variety of good strains, including doubles, is again available. In the last years before the stock changed hands there was a definite deterioration, but all across the temperate world gardeners are holding their breath, hoping that this trend may now be reversed.

### The primrose path of dalliance

Like a number of spring-flowering plants, primroses combine two significant characteristics upon the understanding of which successful

The 'Hose-in-Hose' form of *Primula* 'Wanda' seems as tough as the normal 'Wanda'

propagation depends. The first is that seed is at its most potent when sown fresh; in particular, seed kept indoors in a drawer or, worse still, kept in the greenhouse will deteriorate rapidly. Seed kept in the fridge, in tightly topped glass jars with silica gel added to take up excess moisture, will keep for some time and germinate fairly well.

This leads on to the second requirement. Fresh primrose seed needs moist conditions to germinate rapidly, not hot and not cold, and these are most easily provided by placing the seed pots in a shady frame in spring. This also gives the seedlings longer to mature before the winter than if they are sown in summer when first ripe. Seed can also be sown in January if that is when it arrives from the seed company. In winter the propagator should be employed to keep the seed warm, though certainly not above 68°F, and germination will then depend on how the seed company or seed exchange stored the seed. Seed exchanges in particular may have stored it in ordinary room conditions and by January it may have lost some of its viability.

I sow in $3\frac{1}{2}$in pots of well-drained compost made up of two parts peat-based seed compost and one part fine grit, sharp sand or perlite to be sure of good drainage. After settling the compost, firming gently and smoothing off, the seed is sown thinly and topped with $\frac{1}{4}$in of fine grit then watered in. The pots are then placed in a frame covered with a material such as greenhouse shade netting, which keeps them shaded and protected from downpours and gales, but lets some light through. The seed must *never* dry out.

The seedlings are pricked out 6 x 4 into standard seed trays when the first true leaf is emerging, using the same compost that I use for hellebores: equal parts of sterilized loam, peat, fine grade bark and grit, with the addition of Osmocote for feed. Before the roots become too

entangled the plants are set out or potted into $3\frac{1}{2}$in pots of similar compost. It is essential, as with most potting, not to firm the compost over-enthusiastically but rather to tap the pot briskly on the bench to settle the compost before watering in.

### Primrose, cowslip, oxlip, polyanthus and bunch primrose

I realize that it may seem a little late in the proceedings to make clear the definitions of the names of the plants under discussion. But where could it have been fitted in before now?

The wild yellow primrose is *Primula vulgaris* and it has two wild forms with flowers in a wide range of colours, var. *sibthorpii* and var. *polychroma*, and a white-flowered form from the Majorcan mountains, var. *balearica*. It is also burdened with an enormous range of varieties raised as pot plants, which are too gaudy and lacking in hardiness to merit discussion here. The cowslip, *P. veris*, grows in drier habitats than the primrose and with its dainty yellow bells gathered at the top of its sturdy stems, occasionally it turns up in orange or red or even in double forms. Wild hybrids between these two are not rare, are known botanically as *P. x tommasinii* and are sometimes referred to as false oxlips. This cross is also the origin of the cultivated polyanthus. The bunch primrose, popularized by Miss Jekyll, is a polyanthus with the individual flowers hanging from long stems.

The true oxlip, *P. elatior*, is much more rare, growing only in damp woods in eastern England. It is distinguished from the false oxlip by its flowers hanging down only on one side of the stem; in the false oxlip they hang down on all sides.

There are two other species which demand mention here. *P. juliae* is a very small-growing species from the eastern Caucasus, with a more noticeably running habit and magenta or very rarely white flowers. This has been crossed with other species to give a range of hybrids. Finally *P. megasifolia*, a rare winter-flowering species from the south-eastern corner of the Black Sea. This species has occasionally been crossed with others in this group to create some extraordinary but uncommon hybrids.

### Gallygaskins, hoses-in-hose and Jacks-in-the-green

These are the names given since at least the seventeenth century to unusual, often called anomalous, forms of primroses and polyanthus which have been much prized over the years but never common, for as a result of their unhelpful genetic make-up they do not come true from seed. They sound captivating, exciting, and the Elizabethans and the Victorians cherished them with an affection bordering on avarice. Some are delightful, some are almost ugly – very few are gaudy; almost all are worth growing.

What do all those strange words mean? At this point I had a mind to list them all and give accurate definitions but there seems such uncertainty about what some of them *do* actually mean that this proved more difficult than I had anticipated. 'The moment they appear to be mastered,' says Sitwell, 'confusion begins once more.' Hmmm. Some are certainly clear: a *hose-in-hose* has two duplicate flowers, one inside the other; a *Jack-in-the-green* has a ruff of green, an extended leafy calyx, behind each flower; a *Jackanapes* has an extended calyx like a Jack-in-the-green, but partially streaked or veined the same shade as the flower. Parkinson's description of a *gallygaskin* does not seem to be entirely in agreement with his illustration, but generally the idea of a swollen calyx seems the common feature. The main characteristic of the *Jackanapes-on-horseback* is that it is a polyanthus or cowslip type with a ruff of narrow leaves at the base of the whole head of flowers; there is less agreement about the configuration of the flowers themselves. The *pantaloon* has an enlarged green calyx like a gallygaskin, but striped with the colour of the flower. *Feathers* or *shags* seem entirely extinct but were of cowslip form with each flower in the head, including the calyx, split into slender, yellowish green segments.

To me the hoses-in-hose are the most captivating. I have a hose-in-hose version of 'Wanda' which seems as tough as its less curious progenitor, flowers for months and increases well, but a hose-in-hose oxlip, even more delightful with its pale petticoats, one inside the other, has proved singularly difficult to keep.

### The natural mosaic

With evergreen winter and spring shrubs like daphnes, viburnums, sarcococcas and mahonias as background and the soil transformed from its gravel or its clay to the moisture-retentive yet not at all waterlogged conditions we require, the planting of a living mosaic can begin.

Primroses, of course, in the broadest sense, with choice polyanthus and dainty *P. juliae* alongside the prims of your choice. Snowdrops, the large and lovely 'Atkinsii', its later version 'Lime Tree', and double-headed 'Straffan' or hares'-eared 'Scharlokii'; or the doubles, not the ugly double *nivalis* from supermarkets but those Greatorex doubles with their neat rosette of inner petals, 'Cordelia, 'Ophelia', 'Hippolyta', 'Jacquenetta', plus the extraordinary 'Hill Poë'.

Then scillas and squills: *Scilla mischtschenkoana*, the palest, earliest and the best; *S. siberica*, the darkest blue, plus pushkinias and chionodoxas; muscari need a moment's consideration – they can seed *so* prolifically; dwarf daffodils, starting with 'Cedric Morris' in December; wood anemones, which also boast queer anomalous forms among their many varieties; and violets, though not, here, the smothering *V. soraria*; plus woodsy fritillarias, erythroniums and trilliums, or indeed the other plants mentioned in this chapter.

But it is perhaps wise to apply a little forethought to whether this is to be an evolving mosaic in which the plants are allowed to self-sow, or a more controlled affair in which fat clumps are required and the arrangement not muddled by unpredictable seedlings. If ever-increasing clumps are the aim, rather than scat-terings of seedlings, then double snowdrops, anomalous anemones and double prims should be chosen along with frits, trilliums and hostas, with pushkinias and chionodoxas, single snowdrops, crocus and erythroniums allowed elsewhere. For these will all self-sow and you risk them dominating, but a spectacular mosaic it would surely be and one upon which you may not tread – so make a narrow stone path before planting anything, and tuck your shirt in well before bending to enjoy the colourful marquetry.

# Spring Woodlanders: A Final Choice

Whether it be the Savill Garden at Windsor, the lost wilderness of the Bosnian woods, the gardens and frames at Washfield Nursery in Kent or Jerry Flintoff's perfectly packed yard in Seattle, spring brings such a succession of old friends and sweet surprises, often the more heart-melting for being so fleeting, that following from my main selections I could be in great danger of force-feeding you with an indigestible encyclopaedia of rich fare. So instead here are a few favourite morsels, just three, to melt on your tongue.

## Tiarellas

It's the small and delicate, the dainty and demure which fit here, and where better to start than with tiarellas, all of which except for *T. polyphylla* grow naturally in the United States.

In recent years tiarellas or foamflowers have, in their quiet but forceful way, brought themselves to the attention of gardeners both in Britain and across the Atlantic. They have done this by their undoubted quality as garden plants, but in Britain this increased popularity has resulted partly from the uncovering of an unfortunate deception.

For years we in Britain grew a plant under the name of *Tiarella wherryi*, with neat rounded leaves and elegant spires of fluffy little creamy white flowers. This plant is easy to grow and something of a spreader, but now it turns out to be an impostor. For a few years ago Chris Brickell, who has managed to introduce some wonderful new plants while at the same time keeping the RHS in order as its Director-General and dragging it (sometimes screaming) into the modern age, brought the true plant over from the United States. We saw at once that it was in a different class; to ensure that its distinction is appreciated, it is now known as 'Bronze Beauty' and it's one of my favourite plants – its demure elegance is enchanting in every respect.

The dusky purple, maple-like foliage makes a rolling spring plateau from which rise purple flower stems topped with spires of deep pink buds. These open to fluffy cream flowers with a hint of pink, creating a moody mixture of purple, pink and cream which in dappled, woodsy conditions combines soft solidity and a delicate grace. A single plant makes a tight clump, flowering from spring until well into the summer, and its congested stolons can be pulled into a huge number of tiny pieces which nurserymen root under mist but which you and I can root perfectly well in an unheated propagator at any time from late spring to early autumn.

In America, a quite different revelation has inspired gardeners to take tiarellas more seriously: the introduction by Don Jacobs of several selected forms of native American species. Through his Georgia nursery, Eco Gardens, Don has distributed special forms of American natives in a number of genera including *Heuchera*, *Anemonella* and *Phlox* as well as *Tiarella*. American gardeners are fortunate to have a nurseryman with such a good eye selecting and propagating these plants.

Some of these selections are now beginning to circulate in Britain and I have been especially impressed with 'Eco Maple Leaf', for not only does it have almost jaggedly cut foliage but here in middle England has flowered into August in spite of the attentions of my voracious rabbits. 'Eco Slick Rock' is a most extraordinary little plant; it has small leaves and relatively small flower spikes but sends out galloping runners like a strawberry, which root as they go. 'Glossy' has shining, pale green rounded leaves and 'Eco Red Heart', you might guess, has a rich red, heart-shaped mark at the centre of each leaf.

Only one disadvantage of these plants has come to light in my garden: they all seem noticeably less tolerant of sun and drought than *T. collina*, true *T. wherryi* and the plant we once grew as *T. wherryi* but which turns out to be a hybrid between these two. 'Glossy' in particular collapses suddenly and crisps quickly in too much sun.

In fact there is a second problem: vine weevil seem to have developed a taste for them, and as the blackbirds peck for the grubs just under the foliage whole plants, reduced to frailty by munching, are decimated in seconds.

Don Jacobs has also introduced some impressive hepaticas, especially the marbled 'Eco Blue Harlequin', and anemonellas, although the most stunning of these must be 'Schoeff's Pink', an astonishing fully double-flowered form from Minnesota which hardly seems the same plant as those small and fragile singles. This and other choice anemonellas such as Linc Foster's 'Jade Feather' are certainly more difficult to please than tiarellas; they will thrive and increase steadily in leafy, woodsy conditions, perhaps in a specially made-up bed. They are also a great deal more difficult to propagate, but the brave and the accomplished will take a sharp knife to them early in spring and pot the pieces into a well-drained leafy compost.

To add to its garden virtues, the true *Tiarella wherryi* is a fine plant for the cold, and I do mean cold, greenhouse, where it will start to flower in March and where the early flowers are lovely to cut for posies with snowdrops and early primroses. *Polemonium reptans* is another to which I give this treatment. In the garden, the tiarella sits in front of a good seedling of *Euphorbia amygdaloides* 'Rubra', with *Scilla mischtschenkoana* around about, yet by the time the tentacles of *Geranium pogonanthum* insinuate themselves through the euphorbia in July, the tiarella is still flowering.

'Eco Slick Rock' is more fleeting in flower and more delicate, but its runners clothe a new bed without dominating and smothering as *T. collina* might, and sneak prettily under that wonderful hosta 'Hadspen Heron'. 'Glossy' too has a shorter season but is lovely nestling into the foot of a clump of *Iris setosa*, whose deep green leaves slice through the paler, glossier foliage of the tiarella clump. As I write this in early August, 'Eco Maple Leaf' is enjoying an impressive revival after taking it easy since the end of June. A self-sown *Viola elatior* has conveniently moved in alongside but although the foliage of *Dicentra* 'Bountiful' looks good, this is not a plant famous for its restraint and the tiarella is in mortal danger of suffocation unless remedial action is taken this autumn.

### Dicentra

Well, we must allow him an occasional lapse. Christopher Lloyd's *Foliage Plants* was the first book that made me realize that garden writing could be funny and it still gives me a laugh; but not a word on dicentras. Perhaps he simply dislikes them, for their absence is striking; perhaps he thinks they're too vigorous (seems unlikely); perhaps they don't grow well on his heavy soil (possibly); perhaps their tendency to go yellow in irregular patches annoys him: I think we should be told. I like them not only for their silvery, bluey, sometimes pink-tinted foliage, rather like pale rue, but also for their flowers, dripping from their fleshy stems in pinks and white.

The cheerful insistence of most dicentras on spreading more than is altogether convenient is a mixed blessing, and the writer who said 'propagate by division in spring' was living on another planet; heave them out in handfuls before they smother everything in range seems closer to the real world – and pot up a few pieces for friends while you're about it.

Blooms of Bressingham introduced 'Snowflakes' in 1990. I put one plant under an apple tree, in good woodsy soil, and two years later it has surged outwards and made a clump 5½ft wide by 4½ft deep. It flowers from April to September, like the catalogue says, but seems to produce no more flowers in six months than 'Stuart Boothman', say, does in three. And because it's protected by Plant Breeders' Rights I'm not allowed to pull it to pieces and pot up the bits for sale. And then there's the problem of that scattering of yellow leaves. If I had a large new bed in which I wanted to grow some pugnacious shrubs it might be a good choice, but otherwise...

'Stuart Boothman' may be more restrained but it could never be described as timid. Its purple-tinted, finely cut blue foliage is invaluable, and combined with its dark pink lockets makes a combination that provides an effective plant association in a single plant. But it has no compunction about swamping nascent erythroniums and leaving them for dead, and when it meets the resistance of a mature *Hosta* 'Halcyon' the stream splits and rejoins, leaving the hosta surrounded.

'Bountiful' is less finely cut but its darker flowers are useful, while although the foliage of 'Bacchanal' is green, the flowers are deep glowing red. Jack Elliott's 'Coldham' combines dark flowers and red-tinted blue-grey foliage; 'Lang-

trees' is grey in leaf and almost white in flower. But for white, the rare *D. eximea* 'Alba', correctly 'Snowdrift', is perhaps the one. The foliage may not be finely dissected but it is a good pale rue shade, flat but prettily toothed, and the pure white flowers have a sharp outline. *And* it is much more restrained in its habits.

In quite a different way, there are three other species which can give too much of themselves, but these are climbers. Strictly speaking they should go in a later chapter, for they are distinctly summer-flowering, but if evicted from here I fear they will never find a home. The first many gardeners knew about climbing dicentras was when Bressingham introduced *D. macrocapnos*, a vigorous scrambler with clusters of bright yellow lockets. I planted it under a mahonia on the shady side of a stone wall and it went up the 5ft mahonia, over the wall and down on to the Mediterranean bed on the other side. Then after three years it unaccountably vanished; it has suffered the same fate in the Bressingham catalogue – gone. I would welcome it in the garden again, but it needs strong support, for the mass of shoots will bring down less substantial hosts. It differs from *D. scandens* in its more slender, less showy flowers and seems less permanent.

Propagation of these climbing dicentras is not easy, as although it should be possible from seed, this is hard to come by. Plants can be divided in spring, but it must be done carefully and they should not be treated harshly after such disturbance.

I was hoping to find a place for the next plant, but neither this chapter, nor even this book, is really suitable. For *Adlumia fungosa* flowers in summer, is not a dicentra and is, I have to say, not even a perennial but a biennial. But who am I to let my own rules prevent me from telling you about wonderful plants?

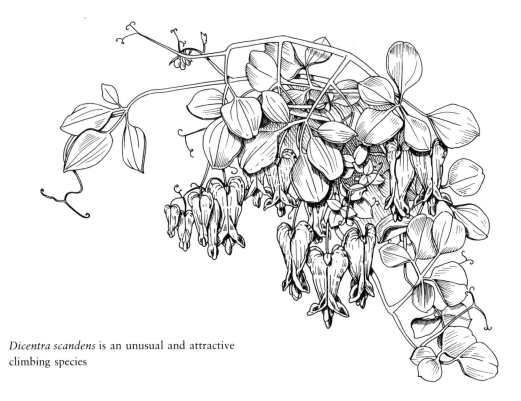

*Dicentra scandens* is an unusual and attractive climbing species

The whole plant – stems, leaves and flowers – is more red than green. It makes a pretty, spiky little mound at first and then in late spring sets off at a great rate, reaching eight, ten feet or more in height by August. The foliage is like that of a dicentra, and in each leaf axil there's a cluster of about a dozen watery purple flowers with dark veins; there is also a wild collected white form which I have not yet seen. It sets seed well and this can be collected from July onwards while later flowers are still at their best. I leave some, and there are always a few self-sown seedlings; one usually seems to tuck itself in the shade behind *Euphorbia* 'Blue Hills' in the Mediterranean bed, which it rapidly swamps.

By complete contrast to all these sprinters, the almost stationary *D. cucullaria* is a model of cool restraint. It differs too in that rather than arch over, the main stems stand upright and the little white flowers hang down like clown's trousers. There is a rumour of a peachy-coloured form found wild in Pennsylvania; it should be introduced to gardens with no further delay. This plant mystifies some because not long after flowering it disappears; but the red tubers are still there underneath. Dormant tubers can be dug up in summer and planted wherever you want them.

In truth, the creeping dicentras are more ground cover plants than choice woodlanders. Where they are best grown rather depends on where you garden; in southern Britain they prefer partial shade and a woodsy soil at least, but the cooler and more moist the climate the more sun they will take.

Like wood anemones they make a cool rolling sea from which shrubs can rise, and are especially good in the middle ground of a shady border, in front of the shrubs and behind the choicer frontal plants, with clumps of mature hellebores placed strategically along the boundary between the two.

I grow them under the shade of old apple trees with islands of hostas like the majestic 'Krossa Regal', *Iris foetidissima* in its various forms and the chunkier pulmonarias like *P. mollis* 'Royal Blue', which seems to let nothing pass. The clumpy *Dicentra cucullaria* is certainly more of a frontal plant.

The climbers can be grown up thickets of hazel sticks, and this is a useful idea in new gardens where shrubs are not sufficiently well developed to provide enough support. At Kew *Adlumia fungosa* used to cover a group of Yakushimanum hybrid rhododendrons completely, although not when they were in flower, and used to set off up a young feathered *Betula utilis*, snapping the lower branches. Mahonias in the *M.* x *media* group like 'Lionel Fortescue' or 'Buckland' have all the necessary strength and resilience, as do the various forms of *Viburnum* x *bodnantense*; in both cases the climbers provide flowers at an extra season.

## Epimediums

How comforting, in this age when we are still being desensitized by African marigolds with names like 'Sumo' and bedding salvias called 'Rambo', that a plant which could never be called flamboyant, spectacular or least of all garish can capture the enthusiasm of so many gardeners.

Much of this explosion of interest in the epimediums or barrenworts has been fuelled by new discoveries in the wild and introductions by botanists such as Martin Rix and Roy Lancaster. As the arrival of new species like *E. acuminatum* and *E. davidii* has started to engage gardeners' interest, more attention has also been given to chance seedlings in gardens and on nurseries, and Elizabeth Strangman at Washfield is screening

seedlings with her usual ruthlessness; so far she has introduced just one new hybrid, 'Enchantress', although more are on the way. The fire is up and it's catching.

Even Professor William Stearn, now over eighty years of age, is working on a new monograph and when this appears it will replace the present standard work – which was published by the very same Professor Stearn in 1936 yet written even earlier in 1932 when he was twenty-one. Mikinori Ogisu, known to some for his rediscovery of *Helleborus thibetanus*, has been extraordinarily diligent in searching out and photographing species in China, some new to science, and supporting Professor Stearn in his work.

The classification of epimediums is certainly a little muddled, but it could be worse. 'These plants have long been confused in gardens,' wrote Professor Stearn sixty years ago; how many times have we heard that? One reason for gardeners'

---

*Opposite*
The yellow skunk cabbage, *Lysichiton americanus*, grows wild from Alaska to California. In wet woods in Washington State the flowers were slightly variable in colour; in gardens, such as here at the University of British Columbia, this brilliant yellow form is most often seen.

*Overleaf (left)*
ABOVE A particularly good gold laced polyanthus. The development of laced polyanthus has been hindered by the restrictive rules of the showbench, and there are few which deviate from this traditional pattern of colouring, seen here in John Fielding's London garden.

BELOW 'Miss Indigo' is a relatively recently introduced double primrose derived from Barnhaven stock, again seen in John Fielding's garden. It is raised in tissue culture and sold in a wide variety of outlets, from specialist nurseries to garage forecourts.

increasing confidence in their familiarity with epimediums might be that in the first volume of the Roger Phillips and Martin Rix book on perennials nearly thirty different species, hybrids and selections are illustrated and described, providing a clear reference for many forms, both widespread and unusual. A second reason is that so far the nurseries which have fostered this increase in interest have taken a highly responsible approach to naming; not, for example, passing off seed-raised plants as clones and being clear that seedlings which may be variable are labelled as such.

But new monographs have a habit of turning our received conventions upside down and Professor Stearn's treatment, over sixty years after his first, will doubtless be no different in that respect. In the meantime it is probably more helpful to classify them in horticultural terms.

The tough, evergreen sorts are, naturally, the best of the epimediums for general planting as ground cover, providing an effective, neat background all the year round, leaving no winter space for weeds to sneak in and being generally the most tolerant of ordinary garden conditions.

In my first year as a gardener at Kew I remem-

---

*Previous page (right)*
'Gareth' is a most impressive new polyanthus which is now becoming available. It was raised by John Fielding from a cross between the well-known 'Guinevere' and a red Cowichan.

*Opposite*
ABOVE *Hosta* 'Halcyon', raised by British plant breeder Eric Smith, is the most popular of the smaller blue-leaved hostas, but in autumn it reveals another side to its character.

BELOW The last October flowers of *Geranium* × *riversleaianum* 'Russell Pritchard' fall over a low wall into the leaves of *Hakonechloa macra* 'Aureola' at the Dell Garden, Bressingham.

ber planting an edging of them around a corner bed of *Berberis wilsoniae* near the Woodland Garden. My naïve credulity was stretched when my new boss, Brian Halliwell, told me that both plants belonged to the same botanical family; the world of botany was curious indeed when an impenetrable spiny shrub was so closely related to what seemed an undistinguished little perennial. But I remembered this unlikely link and noticed too that once it matured hardly a weed penetrated the planting.

*Epimedium perralderianum*, from Algeria, really is good all the year round. In winter the glossy evergreen leaves are invaluable and, like all these epimediums, the contrast between the broad, rather hard leaf surface and the slender wiry stems which support the leaves is one of its most endearing features. Then, finally, when you cut the old leaves down in spring the new season's foliage, tinted with bronze, is coming delicately through, followed by the familiar open spikes of small yellow flowers.

There is a more general point to be made here, for the young growth of some epimediums is undeniably susceptible to icy winds and the time at which the old foliage is removed relative to the season and the state of the new flowering and leafing shoots may need careful balancing.

The flowers of *E. perralderianum* come and go; the bronzed foliage becomes greener, shinier and more crisp and the plant continues its slowly spreading way. This species is one parent of *E.* x *perralchicum*, *E. pinnatum* subsp. *colchicum* being the other. The first record of this marriage was at Wisley and the offspring is so named, 'Wisley'; it combines the bold foliage of the first parent with the larger yellow flowers of the second. More recently this has been elbowed aside by 'Fröhnleiten' from Germany, with slightly larger flowers and foliage which colours brightly in both spring and autumn.

*Epimedium pinnatum* subsp. *colchicum*, one of the most familiar epimediums, is more or less evergreen depending on the garden climate and colours well in the autumn though the yellow flowers are but small. 'None are so dependable for general culture,' said William Robinson. This is a parent not only of *E.* x *perralchicum*, but also of two more hybrids. One is the orange-flowered *E.* x *warleyense*, which was for so long one of the very few epimediums we ever saw, the other is *E.* x *versicolor* which comes in three different forms: 'Versicolor' is cream, with a pinky tinge to the inner petals, the coppery orange form is 'Cupreum', and 'Sulphureum' is its widely grown, clearly self-descriptive variety.

Not only are these evergreen, they are also easy to grow in partial or light shade in a soil which is reasonably rich and not waterlogged. Some are tougher than that, 'imperturbable dense cover, slowly spreading into wide clumps,' says Graham Thomas, and will tolerate brighter and drier conditions, but in these situations choice should perhaps be restricted to forms of *E.* x *versicolor* and *E.* x *perralchicum*.

By contrast, in growing the gems of the genus, some deciduous and some evergreen, we are faced with the necessity of providing conditions which are more carefully considered. They even have names which sound special: *Epimedium acuminatum*, *Epimedium dolichostemon*, *Epimedium leptorrhizum*. But first of all, two which sound rather less exotic but are almost as desirable.

The deciduous *Epimedium grandiflorum* is nevertheless a superb plant in all, and I emphasize all, its guises. For its forms are many and increasing in number, especially when you include its hybrids with other species; most are worth growing, a few are essential. The flowers can be large, though not perhaps quite on the scale of *E. leptorrhizum*, in various reds, purples, pinks

and white, and you can sometimes choose exactly the form you like from nurseries which sell mixed seedlings. The young foliage is also a feature, although it varies a little in the exact shade of its red or purple tinting and the degree of its metallic sheen. Of the many forms around, you can certainly rely, in increasing richness of colour, on 'White Queen', 'Rose Queen', 'Crimson Beauty' and the darkest purple 'Lilafee', a seedling selected by Ernst Pagels.

Then *E. alpinum*, also deciduous, with its superb red-tinted young foliage, slips in here. The tiny yellow and rust flowers may be modest, but that spring foliage is so valuable among late snowdrops and primroses that this tough and easy species is an essential constituent of the woodland mosaic. Even more spectacular for foliage is *E.* x *rubrum*, a hybrid between the two previous species with extraordinarily rich red spring leaves.

Now, temptation... *Epimedium acuminatum* has been introduced twice in recent years, by Mikinori Ogisu and by Roy Lancaster. Seedlings of both have since been distributed, and divisions too; Roy Lancaster's is to be found in many the more gardens. But plants have been lost through not being provided with a bed of moist, humus-rich soil sheltered from both the wind and the fierce midday sun. I grow Roy's form, and the four red arches topped by a neat, four-pointed white roof which make up each delicate flower suggest a most elegant summerhouse or modern garden temple.

Mikinori Ogisu was also responsible not only for introducing but also for discovering *Epimedium dolichostemon* in western Sichuan. Although evergreen, this too will thrive best in special conditions and its graceful stems carry flights of red-eyed white flowers like strange cranes circling in the sky.

*Epimedium leptorrhizum* has flowers like those

eerie crabs from pools in dark limestone caves into which light never seeps; palest pink, almost white, almost translucent, and looking as if they must soon scuttle back to the friendly damp darkness. Finally *Epimedium davidii*, another recent introduction from western Sichuan, this time by Martin Rix, and one of the first of the current influx. Long-spurred, bright yellow flowers are shown off above the neatly toothed foliage; along with *E. acuminatum*, *E. dolichostemon* and *E. leptorrhizum* this species demands a cool, shady spot, leafy soil with good drainage and shelter from fierce winds. Is that too much to provide for such treasures?

Given happy conditions, epimediums will self-sow too. Once you have a few different forms, and providing you are the sort of gardener who is not so authoritarian as to banish every seedling to rotting oblivion, self-sown hybrids will doubtless appear to confuse you just as you thought you had each species clearly in your mind. If you collect the seed, sow it fresh and give it the sort of conditions your intuition will suggest, given my advice on the more tricky species; you can expect a fragile forest of seedlings which will need protection from fierce sun and will, if they have hybridized, provide endless bewilderment and entertainment too.

Split clumps in autumn or early spring but treat them with respect, especially those like *E. davidii* which not only deserve but require it. I confess to having more or less destroyed my large clump by a rather cavalier approach to its division. I apologize.

So, what is it to be? Away with all those lurid rhododendrons under the trees? Move to a garden with a high cover of dappling oaks? A double span net tunnel? Or out with the hammer and saw to build a shade house? Surely, no one can simply go on covering the ground with *Pachysandra terminalis*, a plant whose only role in life

is to be dispatched in a discontented fury to the compost heap to be replaced by something actually worth growing. Like epimediums.

## And finally for spring

A selection of nine spring genera leaves so many aside that even including an extra half-dozen seems to make but little headway with all the others. But bugles must feature for their resilience, their adaptability and their increasing range of colours and forms. Although often thought of as shade-lovers, in conditions which are not too dry bugles make superb softeners of edges even in full sun, creeping out over gravel or paving yet easily reduced should their ambitions exceed your requirements.

*Ajuga reptans* is basically a woodland and hedgerow plant, sometimes growing in unexpectedly boggy patches but almost always with the same green leaves and short spikes of blue flowers. I found a patch of pure white in a small, wet clearing in a Northamptonshire wood, yet there is little hint of the diversity of the garden forms, especially in their foliage.

'Variegata' is one of the prettiest and unfortunately one of the less strong. The leaves are a pale willowy green, edged with white, develop a pinkish tinge in winter, and make the perfect background for the blue spikes. It survives in sun, but remains very sparse. Forms with names like 'Purpurea' and 'Atropurpurea' have long been around, but 'Braunherz' beats them all for richness of colour, a genuine bronze. There are multicoloured forms too, with names like 'Multicolor', 'Tricolor' and 'Rainbow' with pink and white variegations, which you either like or you do not, and a slowly increasing range with other features.

'Silver Shadow' has unusually pale blue flowers but spreads more vigorously than the others, flowering after the main display on those new runners. There is also a pair of unusually large-leaved forms, like young cos lettuces, which have appeared recently and which may be hybrids with *A. geneviensis*. Both have a tendency for their foliage to develop brownish or bronzish tints, 'Catlin's Giant' colours best in sun; 'Jungle Beauty', the largest-leaved of all and also known as 'Jumbo', colours best in winter. Both have good dark blue flowers and are sometimes sold for hanging baskets.

I'm always a little wary of columbines. Knowing their tendency to promiscuity and the resultant unpredictability of their seedlings, and knowing that nurseries sell seedlings raised from open-pollinated seed which may, or may not, be true, I always look at plants and the labels which accompany them with suspicion. Fortunately, in this uncertain world, there are a few to be relied upon; they are so distinctive that they cannot be mistaken.

'Adelaide Addison' is indeed distinctive. The crowded centres of its blue flowers are white but edged in blue, creating a strange rippled effect. Some seedlings come white with a touch of green, and I have also seen a similar form in deep purple and white. Presumably, with careful crossing and selection, other colour combinations could be created.

This seems to have happened with 'Nora Barlow'. Grown since the seventeenth century, when it was known as the rose columbine, it acquired its connection with Charles Darwin's granddaughter rather more recently. Each flower is a tight cluster of slender petals in dark pink, white and green and has tended to come more true from seed than most aquilegias. Then, in the last few years, a mixed strain has appeared in the trade with flowers in the same tight forms but in purples, blues, reds and white, pure colours and bicolours as well as the traditional shade.

Improvements in another old form have also been introduced recently. Plants with green and yellow variegated foliage were known as 'Vervaeneana' but the degree of variegation and the strength of its colouring have varied. Plants with pure yellow leaves turned up occasionally, and the flower colour has varied too, although pale blue and white are the most common.

This variability accounts for the current tendency to gather them together as the Vervaeneana Group, then to give cultivar names to specific selections such as 'Woodside', with variegated foliage and deep blue single flowers, and 'Mellow Yellow', with white or pale blue flowers and soft yellow leaves with no green. The apparent clarity of this system then breaks down entirely, for there are at least two different sources of 'Woodside Mixed' in different ranges of colours, with and without occasional pure yellow plants.

All are good two-season plants in the garden, with the foliage at its best early and again after the flower stems are cut down, with the flowers between. Sown in spring and raised like bedding plants, they also make good foliage plants for tubs.

*Corydalis flexuosa* must be the plant of the decade. First found by Père David in Moupine, now known as Baoxing, in western China in 1865, it grows naturally in the same area as an intoxicating selection of other plants such as *Helleborus thibetanus*, *Fritillaria davidii*, *Davidia involucrata* and *Epimedium davidii*. When people first see a plant, or a picture, or hear of this new blue corydalis their hearts fly in two directions at once: it looks stunning, but aren't blue corydalis difficult to grow? Not this one.

Through the winter there is the neatly divided foliage, which in the form known rather feebly as 'Purple Leaf' is especially attractive, with crisp purple marks at the base of each leaflet, although most plants will show some purple or dusky red colouring in winter. By early spring the foliage has made a low carpet, or in young plants an attractive flat dome, then from March to June comes a constant succession of blue flowers. This is not 'blue' in the sense of murky mauve, and in 'China Blue' particularly, the long slender flowers in the crowded heads are a pure and sparkling shade. Flowering fades in summer, and in dry seasons the foliage may die away entirely, then in autumn growth, and sometimes flowering, recommences.

James Compton, John d'Arcy and Martin Rix introduced three distinct forms in May 1989, having found this species growing in deciduous woods with a *Matteuccia* species, *Tiarella polyphylla*, *Cardiocrinum yunnanense* and *Anemone demissa*. A certain amount of subterfuge was necessary to ensure their introduction. Frustration mounted as they were driven through dappled woods past startling sheets of blue, until the reluctance of their Chinese hosts to pause and allow the team to inspect the plants so inflamed their bladders that an urgent stop was insisted upon, during which time three small pieces of rhizome were secreted in a moss-lined film canister.

On their return each piece was passed to a different nursery and within two years nearly 2,000 plants had been propagated. This is testament not only to the skills of the nurseries but also to the habit of the plant, which increases by stolons from a fleshy rootstock and also produces bulbils in the leaf axils low on the plant.

Each of the three plants proved to be different, so we now have three named forms. 'Purple Leaf' has dark flowers tinted slightly purple when young and leaves which always retain some crimson colour and are the prettiest in winter; 'Père David' has less purple in its leaf, and its sky blue flowers are the largest and almost have a touch of turquoise; 'China Blue' is the tallest plant, with sky blue flowers, sometimes tinged

with green and fading to purple, and dark blotches at the base of the leaflets.

This species has proved unexpectedly amenable in cultivation, thriving in a leafy and well-drained soil in light shade but proving remarkably tolerant of less ideal conditions.

There are now other forms around, and in the States the curiously named 'Blue Panda', with especially light green foliage and sharp blue flowers, is becoming widely distributed. Now that the different forms are producing seedlings it seems probable that they will all soon be muddled, and this exquisite plant may even attain weed status in some gardens before long. But all are beautiful plants which add new habit and a brilliant colour to the woodland mosaic.

There was a time when heucheras or alum roots were grown mainly for their flowers: varieties like 'Bloom's Variety', 'Freedom' and 'Oakington Jewel' were selected for their flowers and some were grown on a large scale for cutting. Recently, there has been a change and the foliage qualities of heucheras have become more appreciated.

The introduction of 'Palace Purple', a bronze-leaved form of *H. micrantha* var. *diversifolia* found at Kew, set everything off, then marbled forms like *H. cylindrica* 'Greenfinch' appeared which were both good for soft flower arrangements and as foliage plants in the border. Mary Ramsdale's 'Rachel', a dwarfer pink-flowered version of 'Palace Purple' promoted by the Hardy Plant Society, was a superb addition, and then from America came the hideous 'Snowstorm' with its white variegation and red flowers. With relief we come back to Piet Oudolf's 'Pewter Moon', a hybrid between 'Greenfinch' and 'Palace Purple' with pewtery silver leaves which are red underneath; unfortunately this is less robust than many heucheras. More recently still, 'Dingle Mint Chocolate' has appeared, with bold chocolate leaves edged with green. More, many more, are being raised in the United States.

Orchids may never be as cheap or as plentiful in supply as humble heucheras, but now that micropropagation of hardy orchids has become a reality they are becoming more widely known and grown. And the very idea of growing hardy orchids in the garden is something which most gardeners find exciting. Many are now less expensive than the best hellebores and so can be treated as genuine garden plants rather than the focus of several of the deadly sins.

*Dactylorhiza foliosa* from Madeira and its taller, more widespread and more luxuriant relation *D. elata* make spectacular plants for leafy conditions, but are among the least fussy of them all and will often thrive in rich border conditions which are not too hot and dry. Their glossy foliage demands, and deserves, protection from slugs, and the resultant richly coloured sumptuous purple spikes have a presence which is impossible to describe.

When I first ordered *Epipactis gigantea* I was alarmed when I opened the packet to find a few dried-up roots whose chances of survival seemed precarious at least. But I planted them in rich but well-drained soil in a made-up bed on the north side of the greenhouse and in their first year there was hardly a leaf to be seen. I sighed, did nothing and pondered my depleted bank balance. In their second year, they looked up and even produced a few flowers; in their third year there was a forest of shoots with brown, red and yellow flowers.

Now I see that Paul Christian is listing eight cypripediums, all raised from seed in the laboratory. He suggests growing them in almost pure silica sand, which seems a harsh way to treat plants which may cost three times as much as this book; however, he knows what he's doing. But although some are only the price of a bottle of good wine, it's perhaps clear that cypripediums do not yet fall within the range of plants we normally refer to as hardy perennials.

*Opposite*
*Rosa* 'Heritage', *Clematis texensis*, *Penstemon* 'Hidcote Pink', *Symphytum grandiflorum*, *Campanula persicifolia* 'George Chiswell', *Geranium* 'Mavis Simpson', *Carex comans*.

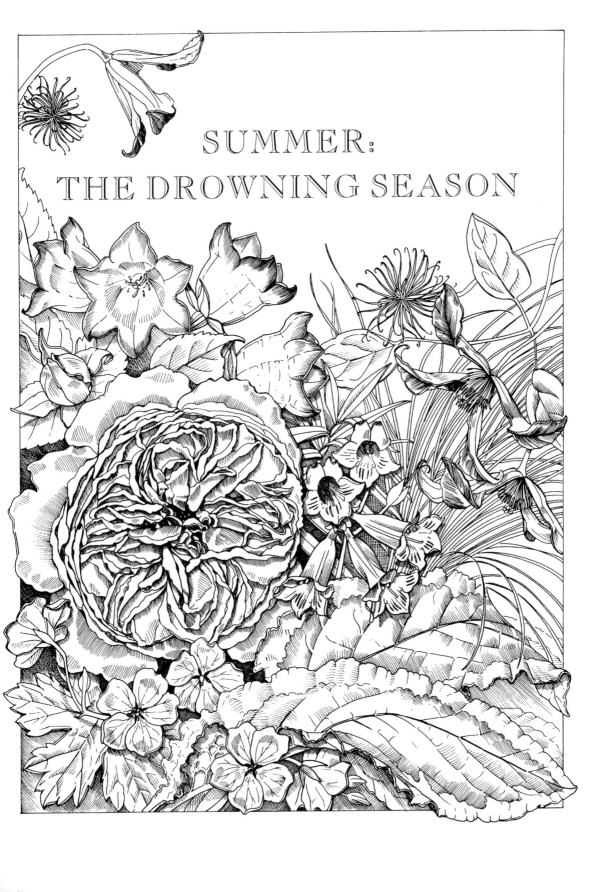

SUMMER:
THE DROWNING SEASON

# SUMMER:
# THE DROWNING SEASON

Summer is the crucial season. It is true that the freshness of spring is special, the soft richness of autumn brings its own warm feeling and in winter, when flowers are so few, every one is precious. But in summer, we drown. Waves of colour sweep over us and however subtle the tones, however inspired and tasteful the combinations of varieties and however ruthlessly we discard the second-rate it can still be overwhelming.

For summer is an ambiguous season. This suffocation beckons us, these draughts of sultry airs and this generosity of colour reach out to us. So, we sink; discrimination deserts us and we drown. Yet, gulping for air, we crave a cool breath, the intimacy of thoughtful choosing, opening to the individual plant whose colour and character momentarily swamp its neighbours. Again this fleeting feeling is soon smothered, for we are drawn now to the expansive sweep of summer's rich ocean, so very many plants and so enticing, and this confidential detail is lost in the tempting expanse of the summer border.

Somehow, in *such* potential profusion, a little more determination is required to focus on any one group, on any one plant – there are just so many distractions. We lurch from drowning in colour and variety to enjoying the individual. In the end there are times, it cannot be left unsaid, when a walk in a green field seems more relaxing, even more pleasurable; this too is unsatisfying.

Alongside this repetitive anxiety runs that final realization (we can deceive ourselves no longer) that some of the care that plants were due earlier in the year they must do without. Now, instead of hoping blindly that there really is still time, we must admit: the asters will not be split, the delphinium seedlings will be sown too late to flower this year, the peonies will be propped up after collapsing rather than staked from the start.

But while there may be regrets, there is such a tide of exciting plants, in such variety and in such profusion that half a dozen gardens, each growing its own particular choice of perennials, could be equally colourful, equally fascinating, equally appealing – and entirely different. So many possibilities in which to sink or from which to choose.

And now I must choose just six groups to discuss.

95

# Campanulas

There are times when I feel in need of a little moral support, when I feel exposed in espousing some extreme opinion; nothing more than a brave fool? This does not prevent me plunging in, and I may find later that one of our past's respected plantspersons, Bowles perhaps, Ellacombe or Louise Beebe Wilder, Farrer or the mad William Robinson, has said it all before – quite probably in the last century or even earlier.

More modern sages, Lloyd of Dixter, Fish of Lambrook, Lacy of New Jersey, may chime in with support – or sometimes flat contradiction, which is fun but unnerving. Just looking for like minds is itself intriguing, and I resigned myself to solitary enthusiasm for *Campanula persicifolia* as one of the finest of all perennials.

### On my own again

I expected Peter Lewis and Margaret Lynch, whose book on campanulas, or bellflowers, is one of the best of recent plant monographs, to be keen; 'refined and good mannered' is their opinion. Mrs Fish helps, but skimps on the praise by only naming it her favourite *herbaceous* campanula. Graham Stuart Thomas is cautious; Bowles, with a tangential view, reports on how much improved is the look of his rock garden after he has removed his jacket and cut down all the old campanula stems; Lloyd delights in the happy combinations its seedlings can create by their unlikely placing.

This is all very well – they like it, I can see that. But my enthusiasm for *Campanula persicifolia* and some of its selected named forms goes beyond that.

First of all the elegant habit: a neat collection of flat rosettes of narrow, deep green leaves, steadily spreading but no more. From these arise, creating a neat sharp angle at ground level, slender vertical stems which carry the flowers. These have a simple precision and a sort of gentleness, except in the rather gross doubles, which is instantly endearing. In the garden their delicacy and apparent vulnerability belies their toughness, in a vase they repay close inspection.

No plant which is so easy to grow that it requires no attention whatsoever can possibly be a favourite – the caring is part of the pleasure – so providing suitable neighbours for shelter and support is important; feeding after the flowers are over and the stems cut back; dividing those rosettes every couple of years to keep them fresh and vigorous; looking carefully for rust pustules and blasting them with propiconazole at the very first sign. I cannot but enjoy looking after such a treasure.

More elegant than phlox, more refined than the delphinium, only the best spotted hellebores can compete – their season and greater variety probably give them the edge.

The simplest forms of this elegant campanula are the blue-flowered and the white, although the blues come in such a range, from just off-white to a deep and shimmering shade. Allow them to self-sow and while blues or whites sometimes dominate, eventually the others will appear.

Larger-flowered single forms come under a number of names. 'Telham Beauty' is a tetraploid with double the normal number of chromosomes and this is reckoned to be the reason for its 4in flowers; it is rarely met with in its true form. 'White Queen' is said to be especially vigorous but there is always the problem of seedling impostors. 'Carillon' is taller than most and said to be resistant to rust, though I have my doubts – not that I've ever even seen it!

Recently a most delightful variation has appeared, a form with white flowers edged with hazy blue. Occasionally appearing in gardens as a seedling, 'George Chiswell', 'Chettle Charm' and 'Blue Edge' are the three names which are current; the first seems to be correct, and the plants to which these names come attached seem identical. The edging is not clear-cut, a neat band, but hazes into white; the effect is delightful. But one step up from these original, simple single-flowered forms comes another group, the ultimate in beauty and simplicity.

To be fair, these cup-and-saucer forms are a little less simple than the wild types, for the calyx, which in the wild species is nothing but a ring of slender green sepals, has become enlarged to make five extra petals. These are fused into a 'saucer' the same colour as the cup, although the lip tends to roll back.

'Hampstead White', more recently also mistakenly named 'Hetty', is the ultimate campanula and at its best is pure white, although Mrs Fish liked it with green tints. The stems are slender and wiry yet strong enough to support the rather heavier flowers; the bells seem slightly larger than in ordinary forms, and there is the added weight of the 'saucer'. The flowering period is a long one; the stems are superb cut for the house and last well; it is vigorous and seems to flower prolifically even if not divided regularly; it even seems to tolerate competition from neighbours better than some – but this may be the blindness of love. 'Gawen' is similar but at 18in is two-thirds of the height of 'Hampstead White'.

Oddly enough, until recently there seems to

have been little in the way of blue cup-and-saucers. Peter Lewis mentions the lilac blue 'Loddon Sarah', Blooms of Bressingham once sold 'The King' but now we have only the recently introduced 'Frank Lawley', a very elegant blue cup-and-saucer which is quickly getting around; the others are cup-in-cup. And here we meet a confusing element in the way these plants are described. Both cup-and-saucers and cup-in-cups (like a hose-in-hose primrose, with one flower inside the other) are sometimes referred to as semi-double, as are plants with three rings of petals (cup-in-cup-in-cup) and also, sometimes, those like 'Flore Pleno' which also have a few small petals cluttering up the centre of the flower.

Surely those like 'Boule de Neige', with row after row of petals, surely those we should call double. Cup-and-saucers and cup-in-cups should be described as such and not described as semi-double, a term which could best be reserved for those in between, especially those with a few remnants of petals in the centre of the flowers.

Now then, these doubles like 'Boule de Neige'. Well, there is no doubt that this is an extraordinary plant. Sometimes called 'Gardenia' and with a touch of the cabbage rose about the individual flowers, the stems need to be a little more stout than normal to support the weighty white flowers, and fortunately they are.

'Boule de Neige' seems to have come under a florist's eye; the doubleness is taken to such an extreme and the flowers are surely a great deal more double than those which the *Botanical Magazine* said in 1798 were pushing the singles out of gardens. Old catalogues are often rather vague in their descriptions – many modern ones are no better of course – but 'Boule d'Argent' which was listed by Kelways in 1912 sounds much the same. But terms like 'very double' and 'full-double' do leave room for doubt.

That Kelways catalogue is full of gems, and

other campanulas of this sort which sound intriguing include 'Candelabre', said to be 'white, with reverse lilac', and 'Marginata', described as 'white; very slightly margined blush'. Blush! And what about 'Phyllis Kelway' – could it really have had *full-double* bells of the largest *size* and of *pale heliotrope* colour of the extremest delicacy of tone' (their italics)?

Keeping these campanulas going is not always easy. Dead-heading not only prevents confusing self-sown seedlings taking over the clump but helps the plant devote its energy to where it needs it most. For those clumps must bulk up, and liquid feeding after dead-heading is very helpful, as is dividing the clumps, regularly, either after flowering or in spring. Rust must be kept at bay; at least it is a different rust from those which attack hollyhocks, roses and mint.

Your favourite forms of *C. persicifolia* are worth treating to pot cultivation, moving them into a cold greenhouse in late autumn for flowering in the late spring. They will not only provide flowers unbattered by gales and rainstorms but they will increase more quickly too.

But wait. There is one form of *C. persicifolia* which should be searched out – and stamped upon whenever it's met; well, I suppose the botanic garden at Kew could be allowed to grow it as a scientific curiosity, and Edinburgh, the Arnold Arboretum and the Brooklyn Botanic Garden, but not another soul. The form I have in mind is so small that a child in a tantrum could finish it off with one petulant stamp of the foot. This is var. *planiflora*, and Lewis and Lynch let themselves down badly when they describe it as 'a rather smart little plant which just misses being first class'.

For me it's as far from first class as the luggage rack on the Trans-Siberian Railway. It reaches just 6in in height, with blue or white flowers the same size as those on plants 3ft high – a preposterous sight. Each plant is squat and dumpy, terms of the ultimate denigration, and the bigger the boot with which it is stamped the more certain will be its demise. Unfortunately, owing to some genetic quirk, self-pollinating some forms of ordinary *C. persicifolia* will give one *planiflora* to every seven tall ones. And then the little midget has the cheek to come true from seed.

The stems are no thinner than those which look so elegant on 3ft plants, and the leaves are oddly congested as if the skin had tightened, making each leaf stiff and slightly curled. The result is not simply ludicrous and entirely at odds with the elegance of the taller forms, but downright ugly. Off with their heads!

### In good company

In Britain we are most fortunate to have five native campanulas, all of which have made the transition from the wild to the garden, without behaving as, or being reviled as, weeds. True, some are best known in double or coloured forms, but from the rare and decreasing biennial *C. patula* to the far more widespread *C. rotundifolia* of roadside verges, all five have found contentment in gardens, and as garden plants have travelled all over the world.

To me, the clustered bellflower, *C. glomerata*, looks better growing on dry limestone banks, as it does not far from where I live, than it does in the garden. This is partly because the form 'Superba' is the one usually grown in gardens and while this is certainly superb in its almost purple colouring, at 2ft in height and with such fat heads of flowers it appears rather coarse and ungainly – and it collapses readily. White forms are pretty but the names are confusing.

In parts of northern England, blue *C. latifolia* takes the place of foxgloves where acid soils give way to limestones and, like foxgloves, white forms are often found among those of the more usual colour. This is a tall plant, reaching over 3ft, with stout stems, a tendency to creep and to produce plenty of seed, and rather striking tubular flowers which burst into a slightly ragged flare at the tip. They sometimes strike out from the stems at quirky angles, some almost facing the sky while others look down.

'Brantwood' is a deep and shimmering blue which is said to come true from seed, however unlikely that sounds. 'Gloaming' is a most wonderful pale smoky blue and perhaps my favourite, and there are at least two whites, one with a dark, slightly bronzy calyx and off-white flowers, the other pure white with no dark tints in the green.

While C. glomerata and its forms are best in full sun, C. latifolia prefers partial shade – not only because it grows better in such situations but because the lack of fierce light helps prolong flowering. Be cautious when raising these forms from seed, as the results can be unpredictable; always divide your best if possible.

At the beginning of this century both 'Burghaltii' and 'Van Houttei' were listed as forms of C. latifolia, but the opinion now is that both are hybrids with C. punctata. Their common feature is dark buds opening to pale flowers. 'Burghaltii' is reckoned to have C. punctata as the seed parent. Its buds are a curious pale metallic purple shade, opening to an unusual greyish mauve; finally the flowers become more blue. I like the unusual and intriguing colour of this variety and in my chalky boulder clay it seems to increase quite steadily. But it flops and does not show itself off well. Conventional staking rarely solves the problem, but creating a more tangled mass of brushwood over the plant works better.

'Van Houttei' is thought to have C. latifolia as its seed parent and the flowers reflect a dominance of this species; most gardeners will know what I mean by campanula blue and that is the colour of the unusually large flowers. This too runs less than C. punctata and has the same need as 'Burghaltii' for discreet support.

There is a much newer hybrid of this sort, 'Kent Belle', which turned up in a batch of C. takesimana seedlings on Washfield Nursery. In its first year it grows to $2\frac{1}{2}$–3ft and then in its second year may reach over 6ft with large, purple-blue bells of waxy, satiny substance. Once established it makes a bold plant for the back of the border, and with a little discreet support follows on stylishly from delphiniums like 'Blue Nile' and 'Fenella'. Alternatively, if new growths are nipped out in spring it will make a bushy 3ft plant in flower from June to September. Although it has a mildly spreading habit, 'Kent Belle' shows no sign of the free running habit of C. takesimana.

Where the blue colouring of 'Kent Belle' came from is something of a mystery, for its presumed other parent, C. latifolia, is not grown at Washfield. Perhaps a customer bought a plant of C. latifolia at another nursery earlier in the day and a Washfield bee slipped through the open car window to gather pollen before visiting C. takesimana.

I first saw C. patula in the garden of photographer John Fielding and admired it at once – it insinuates itself among other low plants so prettily and without dominating. He grew it with 'Constance Finnis' pinks and 'Stapleford Gem' penstemons and it looked delightful; as soon as I find a plant of that lovely little pink I can copy John's idea. Anyway, when I saw seed listed by the American Rock Garden Society I ordered some. I was so pleased with the plants, nestling in front of Iris foetidissima 'Variegata' and among Lamium maculatum 'Red Nancy' with the dwarf form of Lychnis flos-cuculi alongside, that I checked in the seed list to see who had sent in the seed. It turned out to be Peter Lewis, who lives about fifty miles away near Cambridge; the seed had been to the States and back from just up the road.

*Campanula rotundifolia* is one of those plants which I often use as an example when discussing the value of botanical names compared with common names. This dainty little plant is called the bluebell in Scotland, while further south the bluebell is *Hyacinthoides non-scripta*. In the States the bluebell is a mertensia.

It is also a good plant with which to demonstrate how different is the behaviour of a wild species when grown in gardens compared with its more modest growth in its natural habitat. For in gardens *C. rotundifolia* at first makes a

fat and floriferous clump then sets off at a great rate through the border.

This is an oddly unvariable plant – well, that's not strictly true, for over its vast range it does change but within narrow limits; the colour varies very slightly, the leaf shape varies a little, the bells and seed pods are held at a variety of angles. But while it grows throughout the temperate northern hemisphere, America and Europe, it is almost always blue, with only occasional whites which are rare in gardens.

A double form has also turned up but many of the plants listed under this name turn out to be 'Haylodgensis', a double form of the cross between *C. carpatica* and *C. cochlearifolia*, and there is also a form, 'Soldanelloides', with its bell

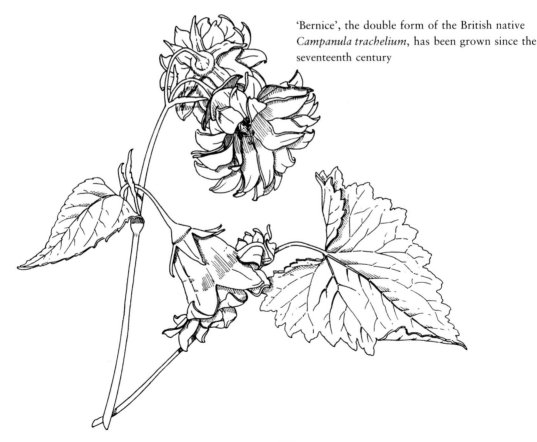

'Bernice', the double form of the British native *Campanula trachelium*, has been grown since the seventeenth century

split into shreds. But no very tall or very short forms, none with very pale or very dark bells, no cup-in-cups or cup-in-saucers...

Seed from a plant in a hedgerow, be it along-side Loch Ness or in Washington State, will produce lovely plants, but caution and a watchful eye are required when seedlings have been planted out.

The final British native campanula is the nettle-leaved bellflower, *C. trachelium*, and surely worth growing, simply for the look of suspicion on a visitor's face when you tell them it is also called bats-in-the-belfry; they clearly believe it's in the gardener's own belfry the bats have taken up residence. Usually the two double forms, 'Alba Flore Pleno' and 'Bernice', are the ones grown, and both have been known since Parkinson's time. The name 'Bernice' became attached to the blue double when Alan Bloom reintroduced it, having received it from Holland in the 1930s.

Both are scarce because they do not set seed and the rootstock is compact and woody; making more by taking a knife or scalpel to it in spring is something that only skilful propagators can bring off successfully and even then the increase in stock is but modest. But such is the demand for these tight doubles that an increasing number of nurseries are listing them, although orders are often met with a regretful 'sold out'.

### Promiscuous punctata

Finally, *C. punctata* and *C. takesimana*. The former is an unpredictable plant and well known. I find it oddly reluctant in old cottagey soil, yet a particularly fine self-sown seedling thrived in a crack in the brick paving – until I tried to lever it out. The long creamy tubular flowers with their flared tips are speckled with red inside. There's an almost white form and also a rather

variable one called 'Rubriflora', which at its best is dusky red with a white picotee edge. I find them all rather susceptible to aphids, unpredictable from seed and easy from division.

Its relation *C. takesimana* was introduced from Korea only relatively recently, and after a short spell on the NCCPG Pink Sheet of rare plants it has rampaged across the country at great speed just as it does through our borders. This species is like a larger and more vigorous form of *C. punctata*, with noticeably glossy leaves and similar, pale pinkish-tinted flowers, spotted red within. *C. takesimana* is bigger, bolder and a great deal more dangerous, even in unimproved soil. But a mature plant, heavily hung with bells, is an impressive sight.

'Elizabeth', another Washfield plant, fits in here. There is some doubt as to whether this is a hybrid between *C. punctata* and *C. takesimana* or simply a good form of the former, although this question may well be resolved by the botanists uniting the two into one species. 'Elizabeth', like 'Kent Belle', turned up among a crop of seedlings, this time of *C. punctata*. This lovely plant has buds which start greenish-white, then become rich red, and finally open to plum pink, darkest at the base and edged with white around the flare of the flower. The flowers are large, heavily pendulous and richly spotted inside with maroon.

It clearly pays to watch the seedlings of these species carefully, especially if you have other campanulas around in the garden. For hybrids have also been found between *C. punctata* and *Symphyandra ossetica*, and this cross produced attractive plants in lilac, misty blue or pink. We should watch for sports too, plants to match *C. garganica* 'Dickson's Gold' but in a different style. Imagine, a 'Kent Belle' with yellow leaves...

101

# Delphiniums

A plant must be exceptional indeed to deserve the accolade 'the true monarch of the herbaceous border'. This may well have been true of delphiniums when it was written in 1933, and in 1949 they were referred to again as 'modern monarchs'. Is there a place for them in the 1990s? Like a politician interviewed at election time, I shall now embark on a circuitous discussion and doubtless arrive at an equivocal answer.

## Wisley and Hidcote

If on a sunny June day you visit the RHS garden at Wisley, there is only one route to take. Emerging from the entrance, you pass the Curator's cottage on your left, up a few steps, past the elliptical beds recently planted with their summer bedding, and turn left. You are then presented with a long vista, herbaceous beds on either side. You are permitted to be distracted occasionally as you pass along the broad grass path between these expansive borders and cross the roadway. There is a hill ahead, so you may rest on the seat behind the hedge to your right.

Then up between the hydrangeas, the lilies and the hostas to the top of the hill. As you approach the top, pause, then slowly crest the brow of the hill and gaze way down across Portsmouth Field. In the dim distance is a haze of pink, blue and white – the sweet pea trial. And alongside, a soldiery of brilliant blue, the delphiniums. Two hundred and fifty plants in over eighty varieties. A gasp is in order.

Now down the steps, and notice the varieties which really gleam, those of such pure and penetrating colouring that they make their neighbours look dowdy – although this is hardly the word for plants simply less brilliant than their neighbours.

Each year the plants in this evolving trial are judged by a joint committee of the Delphinium Society and the RHS, and awards are given. One of the most interesting awards is the single prize, the Stuart Ogg Medal, given to the one variety on trial considered the best of the whole year. This award was initiated in 1982, commemorating the Golden Jubilee of the judging committee, and these are varieties which have been winners: 'Bruce', 'Butterball', 'Gillian Dallas', 'Layla', 'Min', 'Olive Poppleton', 'Sandpiper', 'Shasta', 'Spindrift', 'Summerfield Miranda', 'Sungleam' and 'Thelma Rowe'. They can all still be seen in the trials.

As you approach the bed itself you begin to realize how tall the plants are – most are 6–7ft high, some reach 8ft. And so the first nervous doubts about these monarchs of the true herbaceous border begin to niggle. How can we accommodate such giants? I can assuage these doubts a little by saying that the thoroughness and general high standard of cultivation at Wisley is unlikely to be repeated in the home garden. And it is this old-fashioned level of care which helps create such luxuriant plants.

But even if they never reach more than 6ft in height, fitting them harmoniously into the border requires imagination. At Hidcote Manor in Gloucestershire Lawrence Johnston managed it, and his approach is maintained today by the National Trust. Never one to follow the crowd, Lawrence Johnston used them in his red borders.

They stand at intervals at the back of the border, emerging from behind the early cannas and the bronze-leaved dahlias, and look wonderful. The varieties, like the white-eyed blue 'Starling', are old, but what an idea. When the dead spikes are cut out, the temporary summer plantings grow to fill the gap.

At Hidcote, one mature clump is used at each position. At the Greer Estate in Seattle estate gardener Rick Kyper uses British varieties in large groups behind roses at the entrance to the estate. At Anglesey Abbey, near Cambridge, they feature at intervals along the back of the spectacular herbaceous border around the edge of a D-shaped garden. I was astonished to see that each flowering stem of these broad groups is individually supported by a single slender bamboo cane. The result, of course, is that while the plants look stunning from the bench across the grass, to the extent of overwhelming their neighbours, close up there are only a few days when the canes are not irritatingly obtrusive.

A whole border of delphiniums would be a spectacular sight in the garden for a few weeks, providing there were plenty of beds and borders where other plants could take over for later. But in modern gardens space for such variety of planting is unusual, and this idea of using a single bold clump at intervals at the back of a mixed border is a successful solution. One mature plant, or three close together perhaps, can be slid into the back of a shrub rose border, the flower spikes emerging into their glory and then the remainder of the plant vanishing behind the ever-increasing growth of the roses. But not, please, with a cane to every single spike.

Louise Beebe Wilder seems to agree: 'There is no period of the year when the garden is so exuberantly beautiful as that when there are climbing, tumbling, reaching roses in all directions and spires and spires of Delphiniums gleam-

ing against them or shooting upward like jets of blue flame to touch the fragrant sprays above.' Her roses must have been huge.

As at Hidcote, they can be used in borders of less restricted plantings in the same way, be they colour or season themed – or hardly themed at all.

William Robinson once remarked on 'one of the prettiest effects I have seen among naturalized plants'. It came about as follows: 'Portions of old roots of various kinds had been trimmed off where a bed of these plants was being dug, and in the autumn the refuse had been thrown into a nearby shrubbery, far in amongst the shrubs and tall trees. Here they grew in half-open spaces, which were so far from the margin that they were not dug and were not seen... They were more beautiful than they are in borders or beds, not growing in such stiff tufts, and they mingled with and were relieved by the trees above and the shrubs around.'

This extraordinary passage demands comment. First of all, sticking to the point, this is no more than a random and probably very short-lived version of what was done at Hidcote and of what I am advocating. Secondly, Robinson was obviously so taken with this arrangement that in *The Wild Garden* he described it twice over, in almost identical sentences, in two completely different parts of the book.

But perhaps most interesting of all – why were the gardeners in this unnamed Victorian establishment flinging delphiniums into the shrubbery in the first place? And 'far in amongst the shrubs and tall trees' at that? Was the compost heap too far to barrow them? Or was this some strange Victorian sport?

Well, next time I have some spare delphinium roots I'll drive off to the local forestry plantation and fling them in among the two- or three-year-old saplings. After all, as Louise Beebe Wilder,

Two delphiniums that have been awarded the
RHS Award of Garden Merit, 'Tiddles' (*left*) and
'Lord Butler'

clearly a fan, put it: 'I have never seen Delphiniums poorly placed, they seem so to grace every situation as to make inharmony impossible.' We shall see!

## Monarchs of the border

The continuing work of delphinium breeders, almost all amateurs but many with exhibition rather than garden in mind, ensures a constant succession of new varieties, and, without the facilities to trial them all, deciding which to grow can be difficult. The Stuart Ogg Medal provides one guide, the RHS Award of Garden Merit gives another, and these are the AGM winners to date: 'Blue Dawn', 'Blue Nile', 'Bruce', 'Conspicuous', 'Emily Hawkins', 'Fanfare', 'Faust', 'Fenella', 'Gillian Dallas', Giotto', 'Loch Leven', 'Lord Butler', 'Mighty Atom', 'Rosemary Brock', 'Royal Flush', 'Sandpiper', 'Spindrift', 'Summerfield Miranda', 'Sungleam', Thelma Rowe', 'Tiddles'.

But the variation in opinion among delphinium enthusiasts is striking. For the 1993 Delphinium Society year book, twenty growers were asked to nominate their top six varieties. A total of fifty-four different varieties were chosen, more than half were nominated by only one person, none featured on as many as half the lists and only 'Sandpiper', 'Rosemary Brock' and 'Fenella' featured on more than five out of the twenty lists.

Another survey in 1988 showed how experts differ in their assessments of the same variety. One gave 'Emily Hawkins' 10/10 and another 4/10; one gave 'Mighty Atom' 0/10, while another gave it $9\frac{1}{2}$! My best advice is to visit the trials and judge for yourself, but we cannot all have Wisley on the doorstep. So, being ruthless, here's what I recommend, by colour. Blue: 'Fenella'. Purple: 'Bruce'. Pink: 'Rosemary Brock'. Yellow: 'Sungleam'. White: 'Olive Poppleton'.

## Belladonnas, their flighty offspring

The impression gained by reading different authors on the same subject can be most instructive. And of the two classic books on delphiniums from earlier this century, that by George Phillips published in 1933 is far more infectious in its enthusiasm than that by Frank Bishop in 1949. And Frank Bishop seems only to have included a chapter on the Belladonnas out of duty, while George Phillips did so out of genuine affection for the plants.

'Flowers of unpretentious loveliness,' he wrote, 'blue with the purity of heaven, this dwarf race of a noble and varied genus possesses an infinite and coy loveliness peculiar to itself. Here we have no unfathomable exotics, no stately snobs, but a flower which gazes shyly up at us seeking our affection and appealing to our sense of beauty by the very simplicity of its attraction.'

It makes me want to turn over the whole garden to them, whereas Mr Bishop hardly uses a positive adjective in his whole chapter and struggles to reach 'incomparable for their elegance of beauty' on his second page. Yes, this is a little facetious, but George Phillips makes you feel as if you are about to discover a brother you never knew you had, while Frank Bishop merely reminds you that there are a lot of people in the world.

The original Belladonna arose some time in the mid nineteenth century, probably as the result of a cross between *D. elatum* and *D. grandiflorum*. The first plant from this cross was named *D. belladonna* and during the later years of the century a number of forms were introduced, culminating in 'Persimmon', raised by Kelways in 1897 and Highly Commended by the RHS in 1925.

All these original introductions were sterile, but in 1902 or 1903 Mr G. Gibson of Leeming Bar found three pods of seed on his plants. From the resulting seedlings he named two plants, 'Grandiflora' and 'Mrs G. Gibson'. Modern nurserymen may not be surprised that when 'Mrs G. Gibson' was exhibited at the Shrewsbury Floral Fête of 1905, Dutch nurseryman Mr B. Ruys bought a dozen plants and within a few years was distributing his own seedlings under his own names; one at least of these, the white-eyed dark blue 'Lamartine', is still with us.

Also with us from Ruys' raising are 'Capri' and 'Moerheimii', which are said to have as their rather unlikely origin a single vigorous seedling of a blue-flowered parent. One half of this seedling plant, it is said, produced three spikes of white flowers while the other produced two spikes of blue flowers. The two halves were propagated separately and given those two names. These too are still with us, but it seems far more likely that this curious plant was the result of a lad on the nursery dreaming rather too much of his next, or his last, day off and pricking out two seedlings together into the one pot.

Mr Gibson too went on to raise more varieties, including 'Isis', and Carlile's raised and introduced 'Theodora' and 'Wendy', the latter variety, and its namesake, happily still with us. But in recent years there has been little new work, the imagination and skill of Britain's new generation of delphinium breeders being concentrated upon the Elatum Hybrids.

Perhaps what we need for the garden are Elatum Hybrids in all their wonderful range of colours but without those huge fat spires. Instead the inclination to eliminate or retard the secondary spikes could be reversed and varieties developed which produce a forest of smaller, more slender spikes and no vast spires at all. On reading this the Delphinium Society banishes me from its membership.

In the garden the Belladonnas are invaluable; their clouds of flickering butterflies are entirely without the military bearing of the Elatum Hybrids and so are far more appropriate for chaotic and small-scale borders. They are wonderful foaming around the base of Madonna lilies or to hide the stems of 'New Dawn' roses, and against a yew hedge the white 'Moerheimii' with 'Madame Hardy' roses and spotted white foxgloves would be sensational.

### 'Alice Artindale' and the doubles

It is sad but true that many enthusiasts for hardy plants have become almost as snooty about delphiniums as they are about African marigolds; they are just too colourful. The Belladonnas are beginning to emerge from this neglect, but one which has already done so is 'Alice Artindale', a full double.

Most delphiniums raised recently have been semi-double, fully fertile with about twelve petals and a so-called bee in the centre. But seventy years ago, while there were also singles, there was a small range of full doubles as well.

George Phillips describes the way in which the doubles were split into three groups, although it seems there were not more than a few varieties in each. In the clematis-flowered sorts, the stamens were transformed into pointed petals laid out flat and there was no eye, giving a flower rather like a double clematis. The double-flowered type is rather blowsier in appearance, with the central petals curved inwards towards the centre rather than laid flat. In the ranunculus-flowered form there is row after row of petals, creating a very full, but infertile, flower.

Confusion seems to have reigned over the true form of some of these old sorts. George Phillips, writing in 1933, mentions the following: 'King Bladud', said to be a sport of 'Reverend E. Lascelles', the blue and lavender ranunculus-flowered 'Lady Bath', the double 'Lady Eleanor' in sky blue shot with mauve, the clematis-flowered 'Lady May' in bluish mauve. 'Mrs Foster Cunliffe', in mauve, is listed as a double but described in the text as if ranunculus-flowered. There was also 'Ranunculoides' and the ranunculus-flowered 'Rosette' in purple.

Frank Bishop in 1949 gives the following, saying simply that they are double-flowered: 'Alice Artindale' and 'Glory of Wales' in blue and mauve, and 'Codsall Lad' in dark blue and purple. He also mentions both 'Lady Eleanor' and 'Lady May' but describes them as semi-double.

In recent years only 'Alice Artindale' has been reasonably easy to find; 'Lady Eleanor' has been only occasionally listed. Perhaps those expert amateur breeders of the Delphinium Society, whose skill and imagination have brought them so many of those Stuart Ogg Medals, could broaden their view and recreate some of these old doubles.

### Making more

Seed or cuttings, that is the question? Both are easy. In exceptionally cold, or hot and dry areas seed is probably preferable because the plants will need renewing regularly or you will soon have none. There was a time when the 'Pacific Giants' were superb, but no longer, for they have deteriorated even to the extent that single-flowered plants turn up regularly among the semi-doubles. Collecting seed from whichever delphinium happens to be growing in your or your neighbour's garden will often produce better plants than so-called 'Pacific Giants'.

'Blackmore & Langdon's Strain' is at least harvested from the best named varieties, but in recent years the various 'Southern' series in mixtures and single colours have now surpassed them all. Raised by Len Harrison, in collaboration with Duncan McGlashan in the earlier years, and hand-pollinated every year with parents chosen specifically to produce good off-spring, they are simply the best. But making thousands of hand-pollinations each year is a tiresome business; we need a good open-pollinated strain as the 'Pacific Giants' once were.

Seed sown in the propagator in February will flower well in August and September the same year and an initial screening can take place. But those retained should be well staked, for if the plants are blown over when they first flower they may be lost during the following winter. Seed can also be sown in June for flowering the following year. Plants are best pricked out direct into trays or perhaps better still into $3\frac{1}{2}$in pots. They germinate well if sown no later than the summer after collecting, and rapidly develop into good-sized plants.

Gardens where delphiniums are genuine perennials provide them with a deep, rich but well-drained limy soil, summers which are not too dry and winters which are not too cold – that is, they tend to grow well in Britain, especially the south and west, and the Pacific North West of the USA. Here it pays to grow vegetatively propagated varieties, raised from cuttings; simply because they are better and give the widest choice of colours. It is possible to split the clumps, as those idle Victorian gardeners discovered, but cuttings are easy and give more vigour with less disease.

The most important secret of success concerns the cutting itself. Well, it's not a secret at all, for every book on delphiniums explains it thoroughly. But the only cuttings which will root well are 2–3in long, taken when the shoots first emerge in February or March and before they develop a hollow centre. The shoots should be cut away with a scalpel as close to the woody part of the crown as possible. After tidying up and removing any odd leafy bits at the base, the cuttings can be put individually into 3in pots of compost. John Innes seed compost was once recommended, but a well-drained, peat-based compost is ideal. The pots should simply be stood in an unheated propagating case on the greenhouse bench. As long as the temperature does not fall below about 40°F or rise above about 60°F, rooting should take place quickly. Keep the cuttings damp but not wet, and water in the morning if possible.

If there are no facilities for rooting cuttings in the greenhouse in this way, they can also be rooted on a windowsill. They can be prepared and inserted in the same way but the fluctuating conditions on a windowsill are perhaps better dealt with by a quite different method.

This involves the use of small jars, herb or spice jars about 1–2in across and 3–4in high. Each jar is filled one-third full of washed grit and then two-thirds full of water. One cutting is stood in each jar and the jars lined up on the windowsill. Roots will soon form (how quickly depends on the temperature), and when they are about $\frac{1}{2}$in long they must be potted up. Leave them longer and the brittle roots will be all too easily damaged. Once they are growing away they can be hardened off and then moved outside to a warm sheltered spot before planting out.

They can even be rooted in a cold frame outside, even one made from a bottomless orange box with the top covered by a sheet of glass and a shovelful of grit forked into the soil beneath. But slugs, ever eager to prey on the tender delphinium, could scoff the lot. And because older hollow-stemmed shoots are useless, there is a limited period when cuttings can be taken. If the first batch fails, it may be too late for a second.

## Monarchs of the border?

Pure herbaceous borders are still on the decline and so there are fewer traditional homes for the traditional delphinium. Even in these situations it is not easy to place the modern, most highly developed forms among other contemporary perennials in a harmonious manner; the colours are so vivid, the spikes so tall and fat, that they overdominate. Smaller clumps can be used, but the breeders could help by developing new forms more akin to the Belladonnas.

In mixed borders with shrubs, especially those which also rely on bold tender perennials, it is perhaps a little easier to create the right balance. But here, too, less rigid forms would be valuable and would knit together with neighbouring plants more effectively.

So there is still a place for traditional delphiniums in both herbaceous and mixed borders, although careful thought is perhaps more necessary to ensure that they fit in and do not stand out like the garish bedding plants which most enthusiasts for hardy perennials so determinedly despise. Exciting new forms, which always generate fresh interest, would be welcome.

But instead of varieties which are more informal and less overpowering, we are about to be treated to varieties with scarlet flowers. These may be the culmination of almost a lifetime's breeding work, but their price and the many problems associated with growing them well in the garden are likely to ensure that the cause of the delphinium is put back rather than enhanced.

# Geraniums

For today's plantsman and gardener, the hardy geranium is perhaps the perfect plant. Its popularity is clear from the occasions when it tops polls, the number of nurseries specializing in hardy geraniums, the impressive increase in the number of species and varieties available and the number of National Collections. What is it that makes them so popular?

Cranesbills have that knack of being all things to everyone. Most are easy to grow, and even those which demand a special situation are not so demanding that their preferred conditions are impossible to supply; the challenge is sufficient to demand a little thought, but not so taxing as to regularly defeat us.

They are generally adaptable, and although there is sufficient variation in their natural habitat, from woodland to scree, for them to benefit from us thoughtfully placing the right species in conditions to which it is particularly suited, should we take a chance on less perfect conditions the plants will still do their best to thrive.

Hardy geraniums vary in style so that almost everyone can find plants which appeal – from the new convert from garish bedding to the plant snob who revels in the dowdy and obscure. From the undeniably flamboyant *G. psilostemon* through the colourful but hardly overpowering *G. endressii* forms and hybrids, to the intriguing *G. sinense* and plants like *G. traversii* 'Elegans' which are in themselves delightful associations of flower and foliage; no one dislikes more than a few, and an increasing number of gardeners love them all as indiscriminately as their own children.

Hardy geraniums also have this knack of fitting perfectly into today's mixed borders, which have replaced the formal herbaceous plantings in so many gardens. Their relaxed habit ensures they knit in well with their neighbours and even climb into them. Their effectiveness under and around shrub roses is well known; some, like 'Ann Folkard', will happily scramble to head height through a shrub. On a smaller scale there are modern raised beds where small shrubs, trailers, bulbs, alpines and dwarf Mediterraneans make up an analogous mixed planting, and here too there are hardy geraniums which are at home.

The colour range is not a broad one, but among the blues and pinks, plus white and magenta, some doubles and a few curiosities like the starry *G.* x *oxonianum* 'Thurstonianum', there is enough variation to keep us collecting and comparing.

For those of us with a botanical bent there are taxonomic conundrums to keep us guessing and new, sometimes highly unlikely, hybrids to investigate, while self-sown seedlings might just throw up something new and distinct.

Many are easy to propagate: in autumn or spring they split into more pieces than we need so we can afford to be generous to friends, neighbours and plant sales. Some require more careful propagation so there is scope for the exercise of some skill and again a challenge, but one which it is not impossible to meet.

All these virtues seem to crystallize into the core of their popularity, a combination of variety and adaptability with just enough of a challenge

*Geranium* 'Ann Folkard', raised by a Lincolnshire vicar in the early 1970s

for failure to be a possibility, and enough exceptions to most rules to keep us enthralled. On top of all that one or two are tender, many can be used in flower arranging, there is a scattering of annuals and biennials, some have wonderful foliage, and a few are a real pain.

So it's hardly surprising that while Bressingham Gardens listed only two in their 1939 catalogue (in which they also listed thirty-eight vegetatively propagated lupins) and Perrys listed eighteen at about the same time, fifty years later Bressingham were up to thirty-five and now specialists Axeltree

Nurseries are listing 147. In 1994 hardy geraniums covered nearly ten and a half columns of *The Plant Finder*; in 1993 it was nine. It is clear that while the straightforward attractions of flamboyant lupins have faded, the gentler appeal of hardy geraniums has taken over, even to the extent of their being grown in beds and borders of their own as lupins once were.

### Geranium, geranium, cranesbill and storksbill
Every few years an argument erupts among members of the National Pelargonium and Ger-

anium Society, sometimes detonated mischievously by the gardening press, about what to call the plants in which they have such a dedicated interest. The problem arises, of course, because some of the plants known botanically as *Pelargonium* are frequently referred to as geraniums. Well, I don't care if people prefer to call zonal pelargoniums geraniums, I think the opportunities for confusion are minimal. So while we should usually call a *Pelargonium* a pelargonium, we should not quibble if newcomers to gardening, who may be put off by what they see as irrelevant and pedantic arguments, call a *Pelargonium* a geranium; as long as gardeners grow and enjoy them, they can call them what they like until they become sufficiently interested in the botanical history – which is as follows.

The name geranium has been used across Europe since Dioscorides in the first century AD specifically for plants with beak-like seed capsules; it derives from the Greek *geranion* which itself is derived from *geranos*, meaning a crane. When the African coast was first explored and its plants introduced to Europe, new species were introduced which, although different in some ways, shared this distinctive beak-like seed capsule. At first these were called African geraniums until in 1732 the name *Pelargonium* was proposed for these rather distinct plants. We would probably never have continued calling the African ones geraniums had it not been for Linnaeus. In spite of his contribution in radically simplifying and standardizing botanical names, for which we must all be grateful, he decided that these two types were not really sufficiently distinct to be given separate names – so he called them both *Geranium*.

It seems that even by 1789, when *Pelargonium* and *Erodium* were separated from *Geranium* in *Hortus Kewensis*, the name geranium for the pot plant had rooted in the public mind and that is

where we are today. Oddly enough, while the name *Geranium* was derived from *geranos*, a crane, and we call the plants cranesbills, the name *Erodium* was derived from the Greek *erodios*, a heron – and we call them storksbills. Then the name *Pelargonium* is derived from the Greek *pelargos*, a stork – and we call them geraniums.

For the record the difference, botanically speaking, is this: geraniums have ten fertile stamens, erodiums have ten stamens but only the five inner ones are fertile, while pelargoniums have ten stamens, but only between two and seven are fertile. The flowers of geraniums are entirely regular, those of pelargoniums are noticeably symmetrical either side of a vertical line (mirror images), while those of erodiums are in between. The flowers of geraniums are held in cymes, those of erodiums and pelargoniums in umbels. Pelargoniums also have a nectar tube at the base of each petal.

### Foliage and flowers

Hardy geraniums are primarily flowering plants, so the obvious question is: what about their leaves? It turns out that there are some wonderful foliage plants among the hardy geraniums, although the almost-gold of 'Salome' and *G. sinense* is strongest early in the year; they fade by summer but the colouring in 'Ann Folkard' easily lasts long enough to clash with the flowers. My all time favourite cranesbill, *G. renardii*, has sage-coloured foliage which collects droplets of water and sets off the white, purple-veined flowers perfectly.

From the wilds of the Chatham Islands, home of the almost impossible to grow *Myosotidium hortensia*, comes the far easier *Geranium traversii* 'Elegans', with slightly silky grey leaves to back the pale flowers with their pink marks at the base. The darkest foliage of all belongs to some forms of *G. sessiliflorum* subsp. *novae-zelandiae*, with little scalloped leaves and small pink or white flowers. The form in which the leaves are almost black is well known but there is another with dusky maroon leaves.

Hybrids in this last group are becoming increasingly common, and most are lovely foliage plants. *Geranium traversii* 'Elegans' crossed with *G. sessiliflorum* subsp. *novae-zelandiae* 'Nigricans' has given us 'Stanhoe', raised by Ken Beckett although it may be that the clump sitting unacknowledged and unnamed in the Dell garden at Bressingham for so long originated earlier. But I pre-empt myself.

### The inevitable hybrids

We depend on enthusiastic home gardeners for so many developments in hardy perennials and it is no different here. So while botanist Peter Yeo has made an invaluable contribution in his field and without which the rest of us would be floundering; while David Hibberd has made so many new introductions available through his nursery; while Andrew Norton has revived Mrs Fish's garden, where so many once grew and now again grow; while all this has been going on Alan Bremner from Orkney stands out as a hybridizer of cranesbills, having made over 8,000 individual pollinations (many more by now I'm sure).

About a third of all these pollinations produced at least some seed worth sowing but only 3 per cent, spread across sixty-two different combinations of parents, actually produced flowering plants. So far he has named just thirteen of them. While I've met people who refuse to be believe anyone could be quite *so* crazy, I admit to being deeply impressed by his dedication. But more than that, he has been profoundly generous in explaining to fellow enthusiasts exactly how he went about it. Some plant breeders are so secretive that Faustian pacts seem a distinct possibility. Alan Bremner publishes the details.

Hybrids like *G.* x *oxonianum*, *G.* x *riversleaianum* and *G.* x *magnificum* have been around for some time but, to mention just a few, Alan Bremner has crossed *G. traversii* not only with *G.* x *oxonianum* and *G. lambertii* but also with *G. cinereum* subsp. *subcaulescens* to create an extraordinary little mound of silvery foliage with dark-veined pink flowers; certainly a good new trough plant.

Others have added hybrids more by keeping their eyes open: Peter Yeo has spotted a number in the display beds and standing out ground at the botanic garden at Cambridge, while David Hibberd, Elizabeth Strangman and the National Collection holders have also picked out good seedlings.

Few of these new selections and hybrids have yet reached America, and there are contrasting undercurrents of dissatisfaction, first at the influence of Britain on American gardens generally and second at the time it takes for our plants to arrive there. But if Alan Bremner can do it, surely Americans can do it too – especially as he has explained in such detail how to go about it.

I am particularly taken with hybrids involving *G. traversii* 'Elegans' and *G. sessiliflorum* 'Nigricans'. The former especially is a fine plant in itself and so often seems to lend its hybrids a rim of silvery hairs to their leaves. The latter gives varying degrees of its dark chocolate colour to foliage. 'Stanhoe' was the first hybrid between the two and originated with Ken Beckett in Norfolk; guess who raised the more recent 'Sea Spray' and 'Black Ice'. 'Mavis Simpson', named after the hot-tempered but dedicated gardener in the Duke's Garden at Kew, where this cross between *G. traversii* and *G. endressii* was found, is an especially fine plant to tumble over a rock or large raised bed or to scramble into a tough shrub. Its pale pink flowers are set off well against the silvery foliage. 'Russell Prichard' is an earlier, more flamboyant result of the same cross.

### The magenta myth

The gardening public can be unpredictably stubborn in the face of good advice. Take Mr Bowles, a man of strong opinions if ever there was one. While pointing out, in 1914, that the correct name for *G. armenum* is *G. psilostemon*, he also lays down the law on the subject of the colour of this undeniably striking plant. Having castigated geraniums generally for inheriting 'a pernicious habit of flaunting that awful form of floral original sin, magenta, and rejoicing in its iniquity', he cools a fraction and points out that 'the magnificent black eye of... *G. psilostemon* saves it from being one of the worst astringents of this vision in the whole garden'.

By contrast, that other outspoken writer William Robinson enthuses about this plant in *The English Flower Garden* but describes its flowers only as 'large and handsome' – with no mention of the colour at all. Unless this was nothing but a subterfuge to encourage people to plant it and be startled by the consequences.

However, shortly after *My Garden in Summer* appeared in Britain, Louise Beebe Wilder, an American writer unaccountably almost completely ignored in Britain yet whose writing is full of inspiration, devotes a whole chapter of her *Colour in My Garden* to this controversial colour under the splendid title 'Magenta the Maligned'.

Having pointed out quite rightly that it was not only Mr Bowles who was so vehement on the subject of magenta but also Mrs Earle, not to mention Gertrude Jekyll who called it 'malignant', she goes on to quote Clarence Elliot, who was of a radically different opinion, and introduces an interesting sociological sidelight. 'In some circles it needs as much moral hardihood to say that one likes magenta as it does to confess that one dislikes cold baths.'

Well, I've no sympathy for people who refuse to plant anything but silvery pastels. And I've never willingly had a cold bath or a cold shower in my life – I like a long soak with a good book and preferably a glass of rich red wine. I like magenta and I like *G. psilostemon* in particular, and if that puts me out of line with Mr Bowles, Mrs Earle and Miss Jekyll, at least I have Margery Fish for company. She loved it 'at every stage of the game, from the moment it puts its little pink nose through the soil until it opens its wicked eyes'. Although elsewhere she joins the haranguing of magnificent magenta. And I'm also in the good company of Christopher Lloyd, who has enthused about this same plant though the magenta of *G. nodosum* he condemns as 'villainous and dirty'.

There are two approaches to dealing with this plant, the brazen and the cautious. I have a great urge to plant it with lurid orange kniphofias, *Achillea* 'Gold Plate', scarlet astilbes, fiery crocosmias, lemony anthemis 'E. C. Buxton', white campanulas and purple penstemons in a border of glorious garishness. This delightfully startling planting should, of course, be hidden behind a hedge so visitors come upon it suddenly and you'd need a convenient bench as a comfort for the faint.

The gardeners at Inverewe do not have quite such courage, for although they've taken the plunge with the achillea and the astilbe, they've given in to a cloud of cool blue and silvered santolinas; but then I suppose there must be something in the constitution of the National Trust which forbids it from reducing its visitors to gibbering wrecks.

This is the other approach. Rather than stirring up the storm, the waters are calmed by, in addition to the santolina and the milky blue

*Campanula lactiflora* (not the darker 'Prichard's Variety'), *Artemisia* 'Powis Castle' and *Penstemon* 'Alice Hindley' and 'Stapleford Gem', with *Geranium* x *riversleaianum* 'Russell Prichard' or the rather safer 'Mavis Simpson'.

As Louise Beebe Wilder put it, 'I do not deny that there are poor and wishy-washy tones of magenta and that these are not desirable; but where the colour is frank and pure and used with a right intermingling of green and other soft and friendly hues, there is none more beautiful and distinctive.'

But look, eighty years later the catalogue from Bressingham Gardens still uses the old name of *G. armenum* in promoting 'Bressingham Flair', specially selected for its less strident tone. Neither Mr Bowles on the one count, nor Mrs Wilder on the other, would have approved. Neither do I.

### Beware of the cranesbill

I was quite impressed to see the wild herb robert, *G. robertianum*, used in the spring bedding at Kew one year, and when I found a pale pink-flowered colony by the roadside near Kelso in the Scottish borders I collected a few seeds. So when I came across an absolutely pure white form known as 'Celtic White' I snapped this up too. What a mistake. True, this is a lovely plant with bright green foliage and pure white flowers, not a hint of red in the whole plant, and at first I was delighted to see it seeding about in cracks and crevices.

But when it began to sheet the borders I started to heave it out, and when it began to come up in the seed pots of my alpines, and well before the alpines themselves appeared, it had to go. I still have a little seed put aside and if I ever have a spot where it can do no harm, I can reintroduce it. But in the meantime, this and the Scottish pink one and the ordinary dark version, pretty though they are, will be banished. It will be a

relief to be without that musty smell from the leaves too. It is worth noting, by the way, that these forms do not seem to cross with each other.

Occasionally I subscribe to an overseas seed-collecting expedition, and two species that came from the Alpine Garden Society trip to Japan need a firm hand. As Peter Yeo put it succinctly in his book, '*G. sibiricum* is a weed.' I cannot argue with that. A widely spreading perennial plant reaching a yard across with the occasional tiny white flower on a slender stem, it seems especially adept at smothering other plants and at spreading; like 'Celtic White' it seems at home in mixed borders, scree bed, shady woodsy beds, cracks in paving, the gravel under the greenhouse staging – it gets everywhere. On the continent it even invades lawns... I heave nine-tenths of it out, but always leave a plant or two; after all, four people scaled mountains to bring back the seed. They also brought back *G. wilfordii*, which is in much the same vein although the young leaves are edged in dusky red. It gets the same treatment. If only *G. renardii* was as prolific.

### Splits, cuttings and seed

The propagation of cranesbills varies from the all too easy to the almost impossible. Forms of *G. pratense* seed everywhere and usually produce plants which are a great deal less impressive than those from which they are derived; undistinguished colours and narrow petals giving a star-like flower is a common fault with these seedlings. *Geranium renardii* self-sows enough to provide a few spare plants, but given the prodigious seed production of some species the number of self-sown seedlings can be surprisingly small.

The answer seems to be that geranium seeds have such a hard seed coat that water finds it difficult to penetrate. Sowing natural seed often produces staggered germination, creating that infuriating situation when a few seedlings need pricking out before most have even germinated. The answer, when sowing seed in pots, is to nip the tip off the fatter end of each seed with a scalpel or a pair of nail clippers. Seed of many species then germinates more quickly and more uniformly.

Plants raised from seed can be unpredictable in their appearance, but you can depend on division to give identical plants. Many species develop in such a way that the traditional two-forks-and-pull-to-pieces technique yields plenty of young plants, and these include *G. endressii* and its hybrids, *G. phaeum*, *G. pratense*, *G. psilostemon* and *G. sylvaticum*. As usual with cranesbills there are a few which break the rules, so *G. procurrens* produces runners which root as they go. With *G. macrorrhizum* and its hybrid *G.* x *cantabrigiense* you just snap a piece off in spring and stick it in the ground or in a pot.

But a number of popular cranesbills have very compact rootstocks which can be divided only with difficulty, and these fall into three different camps.

Those based on *G. cinereum* are so promiscuous that although they can be raised from seed the results are extremely unpredictable; these are propagated in one of two ways. Wedge cuttings are taken in early autumn; these consist of a resting bud with a small piece of root attached. They are potted up and stood in a frame or cold greenhouse while they make roots during the winter. Alternatively plants in this group can be raised from autumn root cuttings.

Another group includes such plants as 'Ann Folkard', the various forms of *G. sanguineum* and the delightful *G. sinense*, which can be raised from stem cuttings in spring. These do not always root well, and careful division using a sharp knife and a fungicidal drench is often more dependable.

# Phlox

The *1,001 Best Hardy Plants of 1925* was the title of the catalogue of that year from Thomas Carlile's famous nurseries at Loddon near Twyford in the Thames Valley. A confident title to be sure, and among a mouth-watering choice of superb perennials, the majority of which have now vanished from both gardens and nurseries, were forty varieties of *Phlox paniculata*. Only six of them are now available from any nursery.

By 1939 Amos Perry listed forty-seven varieties, only seven of which are still clinging on, while in the same year Blooms of Bressingham listed sixty-nine varieties, only eleven of which are now available – and none of them from Blooms, who by their spring 1995 catalogue were selling just thirteen different varieties. A sad state of affairs perhaps, although a total of ninety-nine varieties was available from nurseries generally.

## Old and new in border phlox

So the border phlox is not a plant which has disappeared from gardens, it is simply that breeding has continued and new, and in many cases improved, varieties have been selected and introduced. Those figures also reveal how the nursery trade has changed; the large nurseries stock fewer varieties, the rest are spread around an increasing number of small nurseries.

The variety which seems to turn up in most lists of the old era, and which is still offered by a few nurseries, is 'Elizabeth Campbell'. Blooms of Bressingham now grow two plants under this name, one a small-flowered white with a pink eye, rather like *Phlox maculata* 'Omega'; the real thing is rose pink, fading slightly as the flowers

age, with a large white eye, and is a delightful plant. Thomas Carlile called it one of the best pink phloxes and he was right; it is also one of the longest-flowering. I made a note of its impressive display on 25 June and it was still looking good on 19 September; can you ask for more from a phlox?

Some of the others of this vintage are certainly poor. 'Pastorale' has individual flowers which are cup-shaped rather than flat, so they show themselves off poorly, while in 'San Antonio' the petals are floppy and tend to collapse. On the other hand, in addition to 'Elizabeth Campbell' there are some good old varieties still around and I would pick 'Caroline van den Berg', which in spite of its poor foliage has a lovely silky shimmer to its pearly lilac, dark-centred flowers. 'Eventide' is not striking for its white-eyed, lilac pink flowers but for its habit, intensely branched from below half its 3ft; it seems both distinct and valuable. In 'Jules Sandeau' the flowers are bright shimmering candy pink with very pale backs, and as the petal edges turn up slightly the pale shade is revealed.

'Prince of Orange' is one of the most floriferous of all phlox, but a nasty cerisey shade with, thankfully, only the faintest hint of orange, while 'Rijnstroon' has unusually large rose pink flowers with a white ring in the centre; the tubes of the flowers are much darker and show up nicely.

One thing which annoys me greatly about phlox, and which is especially infuriating in small gardens where every inch of leaf counts, is that some varieties lose so much of their foliage by flowering time. The worst offenders in this respect – and offend is exactly what they do – are the white and lilac 'Franz Schubert', the pure white 'Fujiyama', which would otherwise be one of the best of them all, 'Mary Fox' in salmony rose, 'Mother of Pearl' (nothing is entirely perfect), 'Prince of Orange' and pearly purple 'Prospero'. This is due not necessarily to mildew, but to an inherent susceptibility to this condition. 'Rijnstroon' usually remains clothed to the base.

Although many of these old varieties have been superseded, one of those in what you might call middle age stands out. 'Mother of Pearl' was the result of sowing open-pollinated seed at Bressingham in 1954 with the aim of producing a

good, pure, shell pink. What turned up was a plant with very rounded flowers, each petal neatly overlaid on the other with its edge turned up very slightly. Each floret is white with a hazy pale pink zone towards the centre, creating a faint band around the flower.

But those bare stems really do need hiding, and in that wonderful curved border under the wall at Hadspen House in Somerset it grows with the two forms of *Linaria purpurea*, the usual purple and the pink 'Canon Went', together with the pretty fringed pale pink *Sidalcea* 'Elsie Hugh'. A broader association could be built up using other plants in the same border. The dark foliage

'Harlequin' is a variegated phlox with the bonus of reddish leaf tints

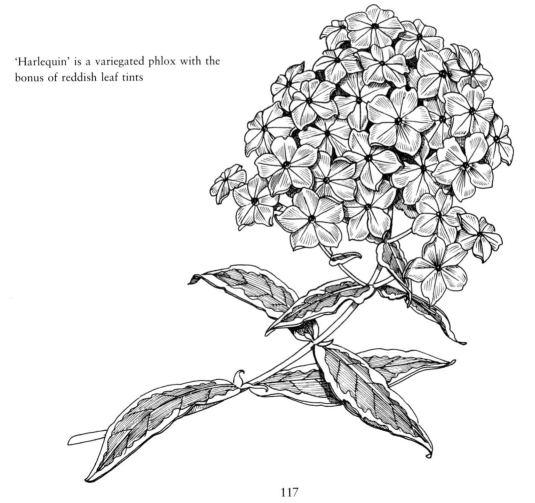

117

of *Aster lateriflorus* 'Horizontalis' fits in well, and the almost black rosettes of *Anthriscus sylvestris* 'Raven's Wing', as long as the flowers are cut out to encourage the leaves, can nestle in front. Black hollyhocks can tower over the back, purple berberis can add substance, *Rehmannia angulata* is perhaps a little too purple but the one known, sadly, as 'Popstar' might just about do; *Origanum laevigatum* 'Hopleys' certainly will. There, a whole summer planting around one of the finest of all border phlox – in addition to the one in which this phlox is actually planted at Hadspen.

Other relatively modern introductions come from Bressingham: no one else seems to have bothered very much with phlox until recently the Woodfield brothers, whose lupins are so stunning, took them up. Those from Bressingham include 'Mary Fox', which, although only 3ft high and with good, cerise-eyed, salmon-tinted, rose pink flowers, seems to be one that stops and starts – it never looks stunning and never looks so terrible as to court the compost heap. 'Eva Callum' came soon after and somehow acquired a completely undeserved reputation for being dwarf; I've been out to measure mine and it has reached an inch under 3ft, with 'Franz Schubert' growing alongside, making nearly a foot less. 'Eva Callum' is, though, a very penetrating shade of sugar pink with a dark eye and is vigorous too. 'Bill Green', raised the same year, is very susceptible to mildew and has all but vanished. The most recent 'Franz Schubert' is neat and a good colour but loses its leaves.

Bressingham also introduced the only two variegated phlox of this type, but I suppose we should not hold that against them. On the other hand... 'Norah Leigh', a sport from 'Border Gem', is one of the most unpleasant plants on this planet, especially when in flower. It is one of those variegated plants with more cream than green in the leaf – and even the green is rather greyish. Somehow the leaves still carry enough chlorophyll to keep the plant struggling on, but, thankfully, not always enough to drag it into flower. For in years when they do appear, the dark-eyed, pale mauve flowers topping those leaves hardly make a harmonious association; time for a sharp exit.

Struggling, I suppose, to be fair, this plant does have a redeeming feature. The leaves are prettily arranged in four neat, vertical, right-angled ranks so the structure of the plant is very striking and attractive. I've seen it looking impressive behind a rolling silvery mound of *Pulmonaria vallarsae* 'Margery Fish', but even this cannot dissuade me from the view that it's an altogether unpleasant plant.

'Harlequin', raised at Bressingham from a batch of open-pollinated seedlings, is more useful. It is much more green in the leaf than 'Norah Leigh' and with the advantage of reddish tints but without that noticeably four-ranked foliage. It has rich pink flowers, dark at the edge with a slight bluish tint on the flat of the petal and an almost scarlet eye. An unreasonable combination.

---

*Opposite*
'Rubicundus' is perhaps the most dramatic of the border penstemons. Although introduced nearly a century ago, it was still good enough to gain an Award of Garden Merit in the recent RHS trial.

*Overleaf (left)*
ABOVE 'Kashmir Pink' is a vibrant form of *Geranium clarkei* selected by Robin White at Blackthorn Nursery in Hampshire. Seen here at Washfield Nursery in Kent, it may flower for only a few weeks but is so spectacular as to be well worth growing anyway.

BELOW *Clematis viticella* 'Madame Julia Correvon' dies back like a herbaceous perennial in cold areas, and here it grows up a tree stump with 'Blushing Bride', John Metcalf's selection of *Lathyrus latifolius*.

In recent years, forms of *Phlox maculata* have become more popular, with their relatively unbranched stems and tall, cylindrical flower heads in contrast to the more congested pyramidal heads of the *P. paniculata* group. They may lack the powerful, and the garish, colours of the more familiar types, but 'Omega' gives me butterflies. Pure white with the faintest blushed eye, it has a cool, confident elegance which says: 'I need nothing, I will be beautiful, even when no one sees me.'

In fact, deep down, 'Omega' and the other border phlox do have needs and these are easy to provide: well-drained but rich soil which does not dry out and plenty of sun. 'Alpha' is pink, 'Miss Lingard' is pure white. There is also 'Delta', introduced, I think, by Vanessa Cook of Stillingfleet Lodge Nursery in Yorkshire, in lilac pink

*Previous page (right)*
ABOVE The white cup-and-saucer flowers of the sublime *Campanula persicifolia* 'Hampstead White', also known as 'Hetty', thrive in this Northamptonshire cottage garden.

BELOW A campanula in an entirely different style – the starry blue flowers of *C. garganica* 'Dickson's Gold' are set against low mounds of bright yellow foliage.

*Opposite*
ABOVE 'Petticoat' is a pretty but undeservedly neglected dwarf phlox, seen growing here at Washfield Nursery. Its split petals suggest it as a form of *Phlox bifida*.

BELOW LEFT Grasses make stimulating plantings without the necessity for other plants. Here, at Burford House in Worcestershire, the blue grey *Helictotrichon sempervirens* is backed by the white-streaked *Miscanthus sinensis* 'Variegatus'.

BELOW RIGHT The Award of Garden Merit winner 'Evelyn' is one of the hardiest of the border penstemons, and here makes a pretty association with the flowers and seed heads of *Clematis integrifolia* 'Hendersonii'.

with a dark eye, and a couple introduced by Georg Arends in Germany in 1918 which seem to have all but vanished.

Among his many other achievements, including the *Astilbe* x *arendsii* hybrids, Georg Arends was responsible for an interesting phlox hybrid which bears his name and which is now being recreated by that imaginative Dutch nurseryman Piet Oudolf. By crossing the creeping woodlander *P. divaricata* with *P. paniculata* to give *P.* x *arendsii*, three varieties, 'Anja', 'Hilde' and 'Suzanne', were raised and released in the 1950s and 1960s, but then disappeared. Now the Hardy Plant Society has rediscovered and reintroduced all three. The slightly straggly 'Lisbeth' appeared much earlier in 1913. It opens a lilac blue and fades towards white; mine has never quite matched its promise and it seems a favourite with my gourmand rabbits. 'Lisbeth' survives still but other earlier varieties like 'Charlotte', 'Katha', 'Louis' and 'Marianne' have gone. All those remaining are fine front-of-border plants and seem to do well on my limy loam; chalk would perhaps be different. I look forward to the new Dutch hybrids.

### Border phlox's friends and relations

Quite a number of smaller, lesser-known species are ripe for discussion here, but I hesitate because the division between alpines and perennials is an indistinct one and all those forms of *P. douglasii* and *P. subulata* which make useful and pretty carpets in drier borders could find themselves under scrutiny. So, as is entirely usual under such circumstances, let prejudice and self-indulgence have its head and let discussion be restricted to just three species, *Phlox bifida*, *P. carolina* and *P. divaricata*.

The ten-point phlox, *P. bifida*, is named from the fact that each of the five petals is so deeply cleft as to give the flower ten points. In the wild, on dry cliffs west of the Appalachians, it makes loose tussocks but in the garden tends to be a little more lax. *Phlox bifida* varies enormously in colour from white, through pale lavender shades to blue and even bicolours like the spectacular 'Petticoat', an unaccountably slow seller it seems.

Then comes *P. carolina*, a perplexing complex of species (according to those who've studied them, not me) under which the invaluable 'Bill Baker' is found, collected by the man himself in North America and flowering from May until September with clean pink stars. This species is also, now, I understand, sufficiently all-encompassing to include the lovely white 'Miss Lingard', which used to be found under *P. maculata*.

All are good front-of-border perennials nestling together happily with short pinks, *Penstemon heterophyllus*, dainty alliums and others which appreciate the same sunshine and good drainage.

Back in the woodsy world is *P. divaricata*. This comes in a wide range of forms, which in the wild can be divided into the lime-loving subsp. *laphamii*, whose petals have no notch, and the more or less lime-hating subsp. *divaricata*, with notched petals; the latter also has a more easterly distribution. The flower colour varies but is usually in the violet to pale lavender range, with occasional albinos, although subsp. *laphamii* tends to have flowers in a richer shade.

Both are at home in shady borders where they root at the nodes as they meander about, the flowering stems standing up about 12in high. A modest revival in this plant has led to Piet Oudolf selecting 'Blue Dreams', in a good mid blue, and 'May Breeze', in blue-tinted white. There is also 'Dirigo Ice' in an icy blue.

Perhaps 'Chattahoochee' fits in here; found by Mrs J. Norman Henry in the valley of that name in northern Florida, the colour is a wonderful rich blue, with a red eye. But seedlings from 'Chattahoochee' come in a range of shades, many

without the dark eye. Linc Foster suspected it to be a form of *P. pilosa*, distinguished by the absence of root nodes, and 'Chattahoochee' has none. At Chelsea in 1994 I spotted 'Chattahoochee Variegated', with neatly cream-edged leaves; it caused a real stir.

### Mildew and eelworm

B. H. B. Symonds-Jeune is credited with great advances in phlox breeding in the 1950s, most of his varieties deriving in part from a discarded seedling he found almost lost in the grass at another nursery. In his book on phlox he describes over two dozen of his own raising, just seven of which are still listed. Many of his seem especially susceptible to powdery mildew, and it is curious that in his chapter on pests and diseases this problem is not discussed or even mentioned. Perhaps he was lucky, perhaps his nursery escaped too much warm dry weather, but his phlox seem not to have been selected with resistance in mind.

However, resistance to powdery mildew is not a constant, as the fungus has a habit of mutating to create new strains which overcome resistance in the host. But the incidence of this unsightly menace can be reduced: ensure the plants have a moisture-retentive, though not waterlogged, soil; never plant under a south wall; if possible plant in tall overhead dappled shade; thin out the stems. Spraying regularly with a suitable fungicide is a wise precaution if you live in a dry part of the country.

Thinning out the weakest stems when they're about 6–9in high not only lets the air through but also prevents spindly flower heads struggling to develop at the expense of the stronger shoots with more potential. In small gardens, there is a simple way of improving the look of those varieties with a tendency to mildew infection and to losing their lower leaves. When the shoots have reached 9–12in high, the tips of those at the front of the clump can be pinched out. The flowers and foliage on the short side shoots which develop will then mask the bare basal stems of the taller shoots behind. Simple. Or of course you could always plant the *maculatas*, whose heavier, glossier leaves seem more reliably mildew resistant.

Then, eelworm. These microscopic animals, usually not more than 2mm long, puncture plant cells and suck out the sap. In phlox they cause a narrowing of the foliage which becomes more pronounced higher up the plant, with the leaf blade reduced to a few frills. Vigour is reduced, flowering is poor; there is no cure. Eelworm usually attacks only *P. paniculata* and its many varieties and fortunately the little beasts only infect the stems, they never penetrate root tissue. As it happens phlox are easy to propagate from root cuttings, so if your plants become infected, propagating fresh stock from root cuttings is the answer.

In early winter dig up your phlox and shake off the soil. Trim off the thicker of the fibrous roots and cut the pieces into 2in lengths. If you require only a few you can lay these pieces of root out flat on a seed tray filled with two parts loam-based compost to one part grit. A little compost is sifted over the roots, a little grit covers the compost, and they are left in a cold frame for the winter. Every 2in piece should have a dormant bud waiting to grow – you can feel them if you run your fingers gently along the root – so each is capable of producing a new plant. In spring the pieces can be potted up individually.

Should you wish to propagate a larger number more easily, the pieces of root can be tied in small bundles with raffia and each bundle inserted vertically into a 5in pot of the compost or in a well-prepared corner outside. When the shoots grow in spring the bundle can be carefully pulled to pieces and the individuals potted up.

In fact, phlox root easily from stem cuttings – even short cuttings taken from flowering shoots in late summer will root. This is just as well, because if you take root cuttings from the variegated varieties the variegation will be lost; all the shoots will be green. So these must be rooted from young basal shoots in spring or from stem cuttings in summer.

Many other phlox, like the *maculata*, x *arendsii* and *divaricata* varieties, which do not suffer from stem eelworm, can be rooted from spring cuttings of new basal shoots. All, of course, can be divided in spring.

# Penstemons

There are two questions which confront those who seek for the truth about penstemons: 'Which are truly hardy?' and 'Which is the true "Sour Grapes"?' Some may add the question: 'Why have so many of the old varieties disappeared?' Now I could simply reply by saying: 'Hardly any', and 'At last, I think I can tell you', and 'I'm not sure they have'; but that would make for a singularly short chapter.

## Which are truly hardy?

The established rule is that the larger the flowers and the broader the leaves, the less likely a variety is to be hardy. So 'Evelyn', with its dainty pink flowers and small narrow leaves, is one of the toughest, while 'Chester Scarlet' would never be classified as reliably hardy. But of course there is more to it than that. In Britain, which for the benefit of American readers is mainly in zone 8, none of the border types are reliably hardy over the whole country. But in the south-west of England, Scotland, Wales and Ireland, just into zone 9, most are hardy in most years, while in the Pennines and the Scottish Highlands, partly in zone 7, none are truly hardy. In the States, gardeners in zone 9 and above should be completely safe. But the truth is that in many areas it depends on the soil, the situation, the wind and the severity of the winter.

The ideal situation for penstemons is in well-drained but rich soil, in a south- and west-facing bed, sheltered from the wind and of course with the benefit of a mild winter. One thing which I find helps enormously is to cut the plants back by about half before the autumn gales. This prevents the wind rocking the roots too much but at the same time allows the base of the plant some protection from snow and ice. The plants are then cut back very hard as they start to grow in spring. In my exposed garden in outer Northamptonshire, on chalky boulder clay, the ones I have found hardiest are 'Evelyn', 'Hidcote Pink', 'Mother of Pearl' and 'Stapleford Gem'; the most tender being 'White Bedder' and 'Chester Scarlet' with the exquisite 'Alice Hindley' in between.

At the RHS trial at Wisley in 1991–2, records were kept of those which survived the winter on the trials field. The toughest, after, it has to be said, the not particularly ferocious winter of 1991/2 proved to be: 'Apple Blossom', 'Burgundy', 'Catherine de la Mare', 'Evelyn', 'George Home', 'Hidcote Pink', 'Pennington Gem', 'Purple Bedder', 'Schoenholzeri' and 'Stapleford Gem'. But even among some of these there were discrepancies; eight entries of 'Schoenholzeri' were sent in by different people and while in most cases all three plants of each entry survived well, in one case none at all made it through the winter. Not one of the eighteen plants of 'Chester Scarlet' made it.

Fortunately, penstemons could hardly be easier to propagate. So in most of the UK the answer to the hardiness problem is simply to take half a dozen cuttings of each variety in late summer and overwinter the young plants in a cold frame or cold greenhouse. Even if the plants in the garden all survive, the display will be stronger and more floriferous if replaced regularly; first- and second-year plants seem to flower best, so

The dwarf blue 'Catherine de la Mare' was one of the stars of the recent RHS penstemon trial

even if old plants survive it sometimes pays to rip them out and replace them anyway.

Almost any above-ground part of a penstemon can be persuaded to root if necessary, even the old woody pieces at the base of the plant. My stock of 'Mother of Pearl' came from one flowering shoot brought to me by a friend in a cottagey bouquet; the stem made two or three two-node cuttings which soon rooted and the following year I had so many plants there were some to spare for a plant sale – they always seem popular.

The tips root the quickest, and the narrower-leaved varieties like 'Evelyn' and 'Stapleford Gem' keep throwing ideal material from the base when flowering away heartily at the top. If they are rooted before the beginning of September pot them up into $3\frac{1}{2}$in pots, if not, leave them until the spring.

### What does 'Sour Grapes' really look like?
To put it mildly, there is something of a mystery surrounding 'Sour Grapes'; but part at least of

the truth seems to be simple. Hilda Davenport-Jones of Washfield Nursery raised 'Sour Grapes' and introduced it in her 1948/9 catalogue, where she described it as 'metallic blue and violet'. But it looks as if Margery Fish created the confusion by listing 'Sour Grapes' in her catalogue and sending out 'Stapleford Gem' by mistake. With penstemons so easy to propagate, the wrong plant became more and more widely distributed over the years.

In her first book, *We Made a Garden*, Mrs Fish had already taken to 'Stapleford Gem': 'It is hard to describe the moonstone tints of "Stapleford Gem". I used to call it "Moonlight" before I discovered its right name and I have met it as "Sour Grapes".' '"Moonstone",' she says later, in *An All the Year Garden*, 'gives you a clue to its colour. It has the same opalescent quality as "Purple Bedder" but on paler lines, and the spikes of rather small flowers are short in proportion to the big leafy bush that it makes.' Ten years after *We Made a Garden* comes *A Flower for Every Day*: 'There is some uncertainty about the name of a good blue penstemon I was recently given. Some people called it "Sour Grapes", others "Unripe Grapes", and both names describe perfectly the soft purple blue shade with an iridescent bloom.' So Mrs Fish alone used five different names for the one plant.

While it could well be that she had the true 'Stapleford Gem' at the start and eventually came by the true 'Sour Grapes', something must have happened in between; perhaps something as simple as an accidentally switched label.

Beth Chatto received the wrong 'Sour Grapes' from Vita Sackville-West and continues to sell it as such for understandably sentimental reasons. But every single entry of 'Sour Grapes' in that RHS penstemon trial turned out to be 'Stapleford Gem'. It seemed as if the true 'Sour Grapes' was lost.

But, as it happens, on the RHS committee which judged the trial was Rose Clay, and when she saw that the true 'Sour Grapes' was not included she showed the committee some flowering shoots from the plant that she had been growing in her Welsh garden for many years. This 'Sour Grapes' was completely different, rich shimmering purple with a little blue towards the base, compared with the much softer colouring of 'Stapleford Gem'. She passed flowering shoots to a couple of nurseries, it was quickly propagated and was on sale, in flower, nine months later.

Fortunately 'Stapleford Gem', which has been doing duty for 'Sour Grapes' all these years, is a fine plant. Its subtle, opalescent colouring makes it a perfect partner for so many plants: soft blue at the base, hazing into lilac towards the mouth and becoming richest at the lips while underneath the tube is pearly lilac with the purple veins in its white throat colouring through. I've grown 'Stapleford Gem' with the pink bells of *Campanula punctata* 'Elizabeth' alongside, where the two can intertwine, and my linaria hybrid, 'Sue', a self-sown hybrid seedling in pink and cream, would be perfect alongside.

### Why have so many old varieties disappeared?

I doubt whether, as is supposed, 190 varieties were sent for trial at Wisley in 1909. It may be true that there were 190 entries, but if the 1992 trial can tell us anything it is that many people will submit the same variety, often under misleading names. Over 100 border penstemons are currently available and while some of these are certainly recent introductions, many are long established; and new names have certainly been applied to old plants. So it may be that fewer have disappeared than we think, and anyway, many old varieties have died out because they were poor plants.

It is true, though, that while 100 seems plenty, there are obvious gaps. But I suspect that many of those which have disappeared were local variants of well-known types, such as the red-with-a-white-throat type typified by 'Rubicundus', raised in 1906. As ever larger and showier flowers were the aim and as these tend to be the least hardy, they were also the most susceptible to extinction.

The confusion has been augmented by a tendency to raise these penstemons from seed. Seed of E. A. Bowles's 'Myddelton Gem', given an RHS Award of Merit in 1909, was listed by Suttons Seeds less than ten years later and it would be remarkable if the seedlings showed absolutely no variation. By contrast individual plants from some mixed strains such as 'Skyline', intended to be grown from seed, have been propagated from cuttings and distributed under the name of the mixture. This is unhelpful.

### 'Continuous colour in the summer garden'

That was Margery Fish's phrase for penstemons, and in spite of the confusion she seems to have created over 'Sour Grapes' she did more than anyone in recent years to popularize them. She grew and sold many varieties, was always keen to try any new one which came to her attention, and used them thoughtfully at East Lambrook. She especially liked to slip young plants in whenever a gap had appeared through an untimely loss, but penstemons are also invaluable for formal plantings, as major players in the mixed border game, and as fillers.

We all have dreams: 'If only I had the time/space/money then I'd make a...' Well, a penstemon border is something that I would like to try. A border filled with penstemons and nothing else would be good, but I always find it hard to leave well alone so I would have difficulty restraining myself from converting it into something more adventurous by the occasional addition of other plants.

I would back the border with a selection of long-season old roses; the colours of the roses and the penstemons fall into a natural harmony. Many of the penstemon shades are certainly muted and blend well, but even the stronger shades are rarely harsh in the same way that old roses are never garish; they seem natural partners.

Better, for their invaluable accomplishment of flowering all summer, the English Roses would be a contemporary alternative. These hybrids between once-flowering old roses and modern roses capture the comfortable elegance and the fragrance of the old roses but with a long season. The silvery shell pink cups of 'Heritage', the stronger tones of 'Bibi Maizoon', the Damask-like 'Mary Rose' and its white sport 'Winchester Cathedral', the dwarf crimson 'The Prince', and the dainty 'Pretty Jessica' – I could go on for pages. They come in a sufficient range of colours and variety of heights to make a sumptuous feature in themselves, and the scent adds one attraction which penstemons lack.

I have to say that I would also be sorely tempted by clematis as companions; whether I would succumb... But viticellas, especially the doubles like the captivating smoky 'Mary Rose', and *texensis* hybrids could scramble through the more substantial roses and be pruned at the same time. While *integrifolias* could snake through the smaller varieties.

In front of the roses would go the penstemons, in ones or groups of three or five depending on the scale of the border. Those I would flatly refuse to do without would be 'Sour Grapes' (true and false), 'Evelyn' and the softer 'Hidcote Pink', the much larger-flowered pink and white 'Beech Park' (sometimes known as 'Barbara Barker'; they're the same though their origins are different), 'Mother of Pearl' and the simply captivating lilac and white 'Alice Hindley'. 'Burgundy', 'Garnet', 'White Bedder' with its chocolate anthers, the bluish purple 'Russian River' – enough!

In front of the penstemons perhaps I would plant those old cuttings-raised verbenas, the ones with the soft colours and the spreading habit like the pink and white 'Silver Anne', cloudy purple-blue 'Loveliness' and the much stronger-coloured, cut-leaved 'Sissinghurst'. Plus the red and pink bicolour 'Pink Parfait', new and very impressive with a lovely scent.

My ambition runs away with me, as you can tell, but faced with only a single sunny corner I could still survive with a plant of 'Heritage' with *Clematis viticella* 'Mary Rose' growing through it, fronted by 'Alice Hindley' and 'Beech Park' penstemons with verbenas 'Loveliness', 'Silver Anne' and perhaps the cut-leaved 'White Cascade' in front. And that little grouping would flower from June to October.

# Grasses

Goodness, it's been a battle. To so many people grass has meant the troublesome but elegant setting in which proper gardening takes place; its only connection with ornamental plants involved autumn pyrotechnics in the pampas and, I might say, roasting unsuspecting hedgehogs in the process.

For so long the feeling persisted that somehow grasses are not real plants, that they cannot respectably be stood alongside hellebores, campanulas and phlox as horticultural equals. The absence of flowers is the problem; those little wisps at the tips of the shoots don't really count, especially as the parts have such funny names. It's difficult enough for many gardeners to remember about anthers and stigmas and carpels without the bother of glumes, lemmas and awns.

But this attitude does little more than expose the conservatism of so many gardeners. All over the country, over the world, mock Sissinghursts are still planted, yet Sissinghurst was begun in the early 1930s; the world has changed beyond recognition since those distant days and although that succession of separate gardens, flaunting or shyly revealing their own individuality, has recently been rechristened 'outdoor rooms', it remains a style which gardeners have clung to for safety – unless beguiled by Alan Bloom's less strongly structured, but clearly defined and controlled island beds.

Personally, I warm to the Sissinghurst/Hidcote approach, but not because I necessarily regard Vita Sackville-West, Lawrence Johnston and the ubiquitous National Trust as examples to revere. My inclination towards hedged enclosures

derives from a need for a certainty of definition. Perhaps Vita, for her own rather different reasons, needed a similar certainty. The tough cookies plant meadow gardens or parks.

But while enclosures or island beds or wild gardens are adopted in gardens all over the temperate world with little awareness of why such an approach is so comfortable, other ideas are also incorporated into a personal setting without any understanding of why or whether they are appropriate, simply because they are generally acceptable. In the same way, bringing us from the level of gardens to that of plants, some plants are taken up by gardeners with an enthusiasm entirely unrelated to any personal aesthetic or the inherent worth of the plant. Many of the hardy geraniums collected so avidly and the hellebores sought so diligently are singularly unremarkable as ornamental plants. Even many that are good are not *that* good.

Conversely, grasses were long abandoned to obscurity because they have had no champion to enhance and solidify their status in the general consciousness. The received view was that hardy geraniums were a 'good thing', grasses were ignored. But the converse of the fact that the 'Sissingcote' approach to garden design is not appropriate for every gardener and every garden is that grasses could be enjoyed by far more people than ever plant or even look at them.

It has been said that the degree to which a gardener appreciates grasses can be taken as an indication of general horticultural awareness. For in a family of so many superficially similar plants, the inclination to distinguish one from another, and then to judge, demonstrates a discrimination, an ability to separate the excellent from the indifferent, at an unusually sophisticated level. And surely it is this fearless implementation of an imaginative discrimination which creates good plantings, an independence of view taking a step forward from history and experience to create art in artifice.

Well, yes, these are fine if not pretentious words. To put it more simply, if you can be bothered to sort out grasses you must have a patience and a clarity of mind which can as successfully be applied to other plants.

### 'A child said What is the grass?'

Take it from me, there are plenty of people who consider themselves grown-up who would do well to ask the same humble question as Walt Whitman. A spirit of genuine curiosity, rather than thoughtless assumption, would quickly reveal that plants with long thin leaves and fiddly flowers with no petals are not necessarily grasses.

There are four groups which are generally thrown together under the heading of 'grasses' but which deserve distinction. The simplest way is this. If a plant has triangular stems filled with pith it must be a sedge – *Carex*, *Cyperus* or *Uncinia*; if it has tiny petals and a cylindrical stem filled with pith it must be a rush, *Juncus*; if it has a cylindrical stem filled with pith and flat leaves with a fringe of white hairs along the leaf edge, then it must be a woodrush, *Luzula*. Otherwise, if it looks like a grass but has none of these features – then it probably is a grass.

Having said all that, so you can refer to them correctly if you wish, from a gardener's point of view they can indeed be treated as one group. For whether you grow them for their flower, their foliage or their general impact, the differences between individual species and varieties are more important than the differences between the general types.

### 'I should make an obtuse-angled, scalene triangle'

Well I might give it a try if I knew what it was. But that is the advice of Christopher Lloyd on the planting of pampas grass, and as the *Shorter Oxford Dictionary* tells me that a scalene triangle is one with unequal sides, I rather presume that he means us to set two pampas grass plants quite close together and one a little further away to one side and back a bit. Seems sensible, now I get the picture, and his notion of filling the gap with *Melianthus major* is a good one too. Though this inspired advice is best not combined with an idea I picked up from a planting in Sheffield Park gardens in Sussex, where pampas is planted by the waterside, its plumes elegantly reflected in the still water and the whole plant floating on its own pool of red maple leaves as they fall in autumn. Pampas + water + melianthus would only guarantee the combination as insistently renewable rather than permanent; wet feet will surely finish off the melianthus every winter. Grow the pampas alone or away from the water's edge.

Or, of course, not at all. This would be my preference unless I had a sufficiently vast estate to allow William Robinson's advice to be followed; he suggests 'a quiet nook where it is sheltered by surrounding vegetation'. This, I like. Imagine a woodland glade: dark conifers provide shelter to the north and the east, your scalene triangle of 'Sunningdale Silver', 10ft high when mature, gleams against the gloom; and in front a grassy clearing, perhaps planted with white crocus for spring, and a seat from which to imbibe the stillness.

And it has to be 'Sunningdale Silver', the most majestic and the purest colour; its stems are also tough enough to withstand autumn gales although the plumes will probably be blown to pieces. Some feel that the shorter 'Pumila' is an improvement but the right spot for it is difficult to find, for it still tends to dominate in the smaller gardens where it's often planted. The fact that its plumes are fatter in proportion to their length is not a recommendation.

Both in Britain and in the States, pampas were once planted in pairs on either side of the front path. Then there was a time when the first thing you did when you moved to a house boasting such silvery sentries was to dig them up. Now I notice they appear occasionally, and singly, in a corner of the front gardens of semi-detached houses, in the angle between the house wall and the dividing fence.

I am content to see them there, and I was also delighted to admire a fine specimen planted in the middle of the central platform of Walton-on-Thames railway station, with trains from London to the West Country flying by on either side. But one of the first things I did on moving to outer Northamptonshire was dig mine up; it would have been an uncomfortable companion for the tiarellas, primroses and autumn gentians which were to go in its bed. My new neighbour and I dug, sweated, heaved and eventually hoisted it into a barrow. We wheeled it next door, planted it, soaked it – and two years later it produced a dozen spikes.

## LBJs

At the Chelsea Flower Show a few years ago I overheard a couple sneering at an especially imaginative stand of sedges and grasses built by Nigel Taylor for Hoecroft Plants: 'More bloody LBJs!' they jeered haughtily, amusing themselves hugely. LBJ is a term used by birdwatchers for those undistinguished little birds, impossible for most of us to tell apart, in which they are not

particularly interested: LBJs – Little Brown Jobs.

This couple were looking in particular at the brown sedges, which they clearly saw as no more interesting than sparrows – which apart from being a terrible slight on the humble sparrow is not entirely fair to the sedges either. For there is no getting away from the fact that although they may be a little more similar than is altogether convenient, they are far from undistinguished.

In at the deep end then, why not start by listing them? *Carex berggrenii, C. buchananii, C. comans* 'Bronze Form', *C. flagillifera* and *C. petriei* seem to be the ones which are available at present. Fortunately *C. berggrenii* is just 2in high so we can hardly confuse it with the others, and as this is one I have not grown let me pass on to the confusing ones.

The distinctive shape of *Carex buchananii* sets it apart; its growth is very upright and strikes outwards to produce a shape like an upturned cone. It reaches $2\frac{1}{2}$ft in height and the thread-like, reddish brown leaves are distinct in being curled like a piglet's tail at the tips. The foliage of the 'Bronze Form' of *C. comans* is a pale, warm, milk chocolate shade and makes a great drooling floppy mass of leaves which may reach 3ft in length and lie on the ground; unless given a haircut occasionally this mass of leafage suffocates its neighbours. *Carex petriei* is like a pink-tinted, straight-tailed, half-size version of *C. buchananii*.

The most confusion is between *C. comans* 'Bronze Form' and *C. flagellifera*, the latter supposedly distinguished by its slightly broader leaves, which are redder or even gingery in colour. To be honest, these two still confuse me. I have bought both and found it hard to tell them apart in the first place when I set them side by side in their pots and examined them, and the latter does not warrant a mention in Roger Grounds's otherwise comprehensive book. Now self-sown carex seedlings are coming up all over the garden but, confusingly, in quite a number of rather similar, chocolatey shades. When the differences are not great and plants appear which seem to be slightly different from either – well, doubts set in.

Most of the sedge fanatics say that these LBJs are best used with blue- and yellow-leaved grasses, and with the sharp, steely blue of *Elymus magellanicus* or the elegant arching stems of *Helictotrichon sempervirens* they are indeed impressive. The similarity of their form and the soft melding of their companionable shades creates a distinct look; lithe and lax, upstanding and overflowing. Like textured brush strokes, the leaf lines in their gingers and bronzy browns, cool grey-blues and ochrish yellows scumble across the dark canvas to build a subtle abstract expression.

But a rolling canvas of moorland shades is not to everyone's taste, and I've been wondering about experimenting with the lower-growing types as ground cover for David Austin's English Roses. The pinks, yellows and whites seem especially suited to a background of brown and bronzy yellow sedges, and icy blue-grey grasses fit in well too. Pearly pink 'Heritage', the much stronger shade of 'Mary Rose', and 'The Countryman' with its strong Portland rose influence look well against both brown sedges and blue grasses; yellow grasses and brown sedges provide an ideal fine-textured wash on which to show off the pure yellow of 'Graham Thomas' and the extraordinary richness of 'Golden Celebration'; plant the white-flowered 'Glamis Castle' and 'Winchester Cathedral', a white sport of 'Mary Rose', against bronze sedge and blue grass.

For browny bronzy tones, underplant these roses with *Carex comans* 'Bronze Form' or *C. buchananii*, perhaps with an occasional acaena such as 'Copper Carpet'; for creams and yellows,

*Molinia caerulea* 'Variegata' has striped leaves, delicate flowers and an elegant, arching habit

both as variegations and pure colours, the water-falls of *Hakonechloa macra* 'Aureola' and *Molinia caerulea* 'Variegata' are especially elegant, and *Carex elata* 'Aurea', *Alopecurus pratensis* 'Variegata' plus the indispensable *Milium effusum* 'Aureum' should fit in well; for greys and steely blues, *Koeleria glauca*, *Sesleria caerulea*, fescues of course (as yet, I cannot see a way through the increasing number of their apparently indistinguishable forms), and acaenas like 'Blue Haze' for the front.

### A top five

Now I realize that some gardeners will remain unconvinced, those stubborn sceptics perhaps more contentedly wedded to their amiable old phlox and delphiniums and steadfastly eschewing a fling with grass. Fortunately, my morality of gardening, such as it is, remains a liberal one and I would no more exclude a whole family or genus of plants from the garden simply because I was locked in the taste of an earlier generation than I would continue to eat roast beef and Yorkshire pudding and eschew pasta.

There is certainly some truth in the notion that the exclusion of certain groups of plants owes much to the poverty of the gardener's imagination – whether this be exhibited as a desire to use certain types of plant creatively coupled with a sad inability to carry it off, or the simple refusal to consider anything but the familiar and the safe.

So by way of temptation, by way of a gentle introduction to using grasses effectively, let me pick half a top ten, a top five of my favourites, and suggest how they might be used. Try them, please.

First of all, for that sharp, icy, grey-blue shade like slender slivers of glass, an *Agropyron*; but at once we have a problem. The plant I have in mind is one I grew from seed obtained from the seed exchange of the American Rock Garden Club as *A. scabrum*. Compared with the plant I already grew as *A. pubiflorum*, the four seedlings I raised produced stems and leaves which grew noticeably flat on the ground and this habit of making carpeting rather than upright growth caught my attention. I gave one seedling to Elizabeth Strangman at Washfield Nursery and was then surprised to discover that my three faded away during the harsh winter that followed.

The plant at Washfield flourished, soon started to develop quite a different habit and was clearly determined not to lie flat; it eventually grew 3ft tall and much more like my *A. pubiflorum*. This pillar of soft Jaeger blue is the plant I pick as one of my top five. It is now known as *Elymus magellanicus*, a name used to cover all the tall, upright forms.

In a blue planting it would look lovely erupting from behind a medium-sized blue hosta such as 'Bressingham Blue', or one such as 'Royal Standard' which develops good autumn tints. This green-leaved 'Royal Standard', with its white flowers in late summer, is a fine companion and in front *Lamium maculatum* 'White Nancy' is just right. This grass also sets off border aconitums well, both the pale *A. carmichaelii* 'Arendsii' and the darker 'Kelmscott', Belladonna delphiniums in white and blues, white 'Fujiyama' phlox for a chilly look, and perhaps with those wonderful new aster hybrids, 'Ochtendgloren' in pink and 'Herfstweelde' in blue.

My next choice would be *Carex buchananii*, and this is probably the one that will tax the tolerance more greatly than my other selections. This is not because of its colour, which is a fiery ginger-rust, but because of its habit and the length of its leaves. Having just been out with a tape I can tell you that by September some leaves have reached 4ft in length, making a plant 2ft high and 5ft across. If the growth was as dense as that of *C. comans*, it would be an awkward plant indeed but this is not so. Its habit is mostly arching, the leaves hanging in languid curves, and is sufficiently open for other plants to amble through. Today *Geranium versicolor* is doing just that, still opening its prettily veined flowers, but its growth is less dense than that of *G.* x *oxonianum*, of which it is a parent, so it meanders in and out without dominating. I did wonder if the pink-on-grey *G.* x *riversleaianum* 'Mavis Simpson' would be a good companion but it too has rather suffocating growth.

I grow the sedge with *Narcissus* 'Jumblie' around the base for the spring and with *Mertensia virginiaca* for its soft foliage and pendulous clear blue bells, yellow-and-green-leaved *Symphytum* 'Goldsmith', the whitest hybrid bergenia, 'Beethoven', and then *Euphorbia amygdaloides* 'Rubra' to pick up the dark hints again.

'Alboaurea' and 'Aureola', two different forms of *Hakonechloa macra*, have become rather muddled and my suspicion is that whichever name you buy, you usually get 'Aureola'. 'Alboaurea' has weaker growth, accounted for by the almost total lack of a green stripe among the white and yellow in the leaf. 'Aureola' is brighter yellow, with fine green lines and altogether stronger.

This plant has three great assets. First of all its colour, which is bright without being garish. Second its wonderful growth habit, like a yellow waterfall running over a rock. And if you ask for more, I give you its increasingly reddish then rusty colouring in autumn.

However, it is undeniably slow to clump up sufficiently to make an impact and perhaps this is an unforeseen aspect of its success as a partner for *Hosta* 'Halcyon'; both take a while to settle down and mature into their true character. As the clump increases, and you can speed things up by planting three together in a tight triangle,

you must beware of letting more vigorous neighbours smother it or it will fade away. It needs good, rich soil and light shade such as that cast by a high leaf canopy.

You might think a mat of *Lysimachia nummularia* 'Aurea' running underneath would make too sunny a combination, so you could consider an acaena such as 'Blue Haze' or the rounded, scalloped leaves of *Geranium sessiliflorum* 'Nigricans', their bronze setting off the grass well. Mature clumps can clash elbows with more forceful neighbours and *Coreopsis verticillata* 'Grandiflora' looks fine coming up behind; the half-sized 'Zagreb' at 12–15in will send its flowering stems through the shower of yellow leaves.

Perhaps the most stunning of all the smaller grasses is the Japanese blood grass, *Imperata cylindrica* 'Rubra'. This is an indispensable autumn plant and whatever else you do I insist you plant it where you can look at it with the sun behind it. This creates such a warm and alluring glow, like a quiescent and friendly furnace, that you will need it near. It prefers a rich soil that is well drained, also a little shade, and is not hardy in colder areas. Like the hakonechloa, this is a plant which dislikes smothering, as I have found to my cost, partly because it is late into leaf and earlier, stronger neighbours can easily muscle into its space.

Choose neighbours with care, therefore. Tight clump-forming plants which keep their place or delicate mats are ideal. If the soil is right the yellow-leaved lysimachia *might* again do. The white-edged grey foliage of *Ajuga reptans* 'Variegata', weaker in growth than other bugles, makes a neat mat, while in Dick Meyer's warm garden in Columbus, Ohio, the silvery undercarpet is provided by *Orostachys furusei*, through which the blood grass stands like red-plumed sentries.

All these grasses and sedges are easy to propagate by division in spring. The carex, hakonechloa and blood grass can be split into quite small pieces if never allowed to dry out and if provided with a cosy environment in which to grow on until rooted into their pots well. Splitting in the autumn is less reliable, although if simply dividing a large clump into good-sized pieces the risk is rather less. The elymus can also be increased from seed sown in spring which germinates promptly with little trouble.

Indeed the carex and the elymus may well self-sow, although *C. buchananii* seems more circumspect than does the brown form of *C. comans*. With me this latter has made an attractive self-sown trio with the blue agropyron and my final choice, the soft yellow of *Milium effusum* 'Aureum', all having sown themselves into a multicoloured clump. But you must be aware of self-sown *comans* forms, they bulk up so quickly and can cause problems in the wrong spot.

Bowles's golden grass, *Milium effusum* 'Aureum', is the last of my five and a real favourite. It seems to have that extraordinary capacity to place its self-sown seedlings where they fit in naturally with their neighbours. And although a perennial, and a resilient one, it never smothers. In spite of its name the leaves are not gold, they are yellow and soft in texture as well as in tone. This is a plant which may put itself anywhere from full sun to full shade, but in shade is liable to become rather green, as it does by September anyway. You can split it, but why bother when seedlings come true?

This invaluable capacity to slide self-sown seedlings into just the right spot has left me with some happy groupings. A number of seedlings slipped in behind a mature clump of *Hosta* 'Thomas Hogg', and although relatively invisible at first, when the light, airy sprays of tiny yellow flowers danced above the variegated hosta it

looked wonderful. A seedling put itself in among *Dicentra* 'Stuart Boothman' with *Viola elatior* and that too looked just right, for the one thing that seems not to suit this delicate and independent grass is to be planted, as is sometimes suggested, in a group at the front of the border. It needs to be among things, to create soft surprises; and on those rare occasions when the surprises turn out to be harsh, you can simply pull the seedlings out.

### Be warned

Finally, a mention of a few horrors – and there is no denying that there are some. Three should never be allowed in a small garden and should probably be outlawed by a resolution from the United Nations.

*Carex riparia* 'Variegata' is a sneaky plant. It sounds good: slender, upright foliage with broad white stripes and black flower heads; it even looks good, at first. But when shoots start coming through the path 3ft from where you planted it (with no sign of it in between), and when plain green shoots start to emerge through hostas so slow you wonder if they are ever going to mature, it becomes clear that a cancerous catastrophe is close at hand. You must repent your yielding to temptation and ignoring the wise counsel found in books like this. You must dig up the border and have it out.

Equally sneaky but in an entirely different way is a form of the much appreciated *Deschampsia caespitosa* known as 'Fairy's Joke'. As far as I can tell this is no more than an especially deceptive selling name for var. *vivipara*. The plant itself is attractive enough, reaching 4ft with dark green arching leaves and airy sprays of green flowers which look very elegant at the back of the border leaning forward protectively over smaller treasures. But.

In this form, the stems do not carry flowers

but tiny plantlets, hence the name var. *vivipara* meaning 'giving birth to live young'. This is all very fine you might think, it must be an easy plant to propagate. Dead right, but it takes a particularly slippery nurseryman to discover a plant which any idiot can propagate by the trayful then give it a sentimental name in the hope of duping unsuspecting gardeners. 'Fairy's Joke'? The only person smiling must be the nurseryman who thought up this wheeze and he must be in fits. The joke is lost on his trusting customers.

For those long stems lined with leafy little plantlets catch far more rain than ordinary flower heads so they soon bend over under the weight of water; and where they touch the soil, they root. Soon you have a veritable forest of young plants racing to repeat the procedure and your more delicate plants are in severe danger. If you have a plant, cut off the flowering stems and burn them; then dig up the rest of the plant and burn that too.

Finally a plant which could cause problems for fans of Gertrude Jekyll. Lime grass, *Elymus arenarius*, grows on sand dunes around the British coast and makes very attractive broad patches of grey-blue foliage. Miss Jekyll recognized the value of its cool colouring and planted it at the ends of her 200ft border at Munstead Wood, with grey santolina, sea kale and stachys and also with double soapwort. I cannot quarrel with the fact that these would indeed make attractive colour combinations – in the short term. But this is a rampager of the worst kind, couch grass on roller skates. Graham Thomas puts it mildly when he says it 'cannot be recommended except for large waste areas... Gertrude Jekyll was very trusting when she recommended it; in her light soil it would be highly invasive... Her garden staff must have hated it.' Perhaps she would have used *E. magellanicus* had it been available at the turn of the century.

Should you spot a grey haze in the distance while walking a coastal path, turn your head to admire the sea view in case you be tempted.

# Summer – The Drowning Season: A Final Choice

The profusion of summer plants is such that having chosen just six summer groups to examine in detail, so very many are left to this final discussion; and every time I examine the list I add more. But it is exactly this abundance which makes the fascination of perennials so limitless, our gardens so varied and allows every gardener to develop and express a personal taste.

## Clematis

Many gardeners are distracted from the virtues of the herbaceous clematis species by the flamboyance, familiarity and sheer numbers of the large-flowered climbing hybrids. A few are simply unaware of their existence and would perceive the herbaceous species only as some irrelevant oddity were they ever to come to their attention. Others glimpse them only in the periphery of their horticultural vision, never fixing their gaze directly upon them; they linger on the outer fringes, in a wilderness from which they may, or more often may not, be rescued.

But were it not for their dazzling relations, prodigies leaving little attention for less blatant siblings, herbaceous clematis would be inspected and assessed simply on the strength of their own virtues – which would be seen to be considerable.

The herbaceous clematis break down into three rather mixed geographical groups: the Europeans, the Asiatics and the Americans. The species from Europe are the best known, especially *C. integrifolia*. Its slender stems reach about 2–3ft in height and need shrubs for support; plant the clematis alongside a potentilla, a penstemon or a hebe, even in the same planting hole, and it will surge up through the middle before falling elegantly around the outside. Then the stem of each individual flower arches over so that it nods, its four narrow fleshy petals twisted, reflexed or with the sides rolled.

Some forms of *C. integrifolia* are less than prolific in flower; a single spark of blue at the end of a long stem is not enough. The flowers of 'Hendersonii' are larger than most, and there is a striking Oxford and Cambridge contrast in the blues of the edge and centre of each petal. It's prolific and long-flowering, with silvery seedheads and flowers on the plant at the same time.

'Olgae' has strikingly reflexed petals, rather than twisted as in 'Hendersonii'. They are deep blue at the base, fading to a soft, downy, almost waxy blue, but there is more than simply a solitary flower at the end of each shoot. For there are branches from low down, then each subsidiary shoot produces not only a flower at the tip, but two flowers at each leaf joint. The result is far more flowers over a far longer period.

These are both old forms, as is the bright candy pink var. *rosea*. 'Pastel Blue', one of a number of newer selections raised by Devon nurseryman Barry Fretwell, has very large pale blue flowers, ribbed in slightly darker purple and a slight, sweet, lily-of-the-valley smell. 'Tapestry' is also his, with mahogany red buds with white hairs along the ribs, mature cerise flowers and silky seed-heads all at once.

The other European is *C. recta*. In its usual form this can be difficult to manage, as it makes a great mass of growth 6ft in height and although foamed in scented white stars, the whole plant is distinctly collapsible. Barry Fretwell has selected 'Peveril' at a more manageable 3ft, and there is also 'Purpurea'.

'Purpurea' is an example of a plant raised from seed in the expectation, or hope, that it will be a good colour and sold without careful assessment and selection. Only by vegetative propagation can you be sure of a good form. So buying by mail order is foolhardy; you may end up with a poor greenish-leaved plant. For although they all turn green in the end, at its best the young shoots of 'Purpurea' are a rich, deep purple. Perhaps Barry Fretwell is working on a dwarf dark-leaved variety. There was once a double form, said to have white buttons like *Ranunculus aconitifolius* 'Flore Pleno', but it seems to have been lost.

Two Asiatic species in particular are worth mentioning. *Clematis fusca* is a little gem, though a very variable one. Its stiff growth may reach 8ft and remain woody, or it may peak at 18in and die back. The flowers are like tiny, hairy urns and come in a range of browns, purples and golds; there are number of subspecies in Korea and Japan and while none are flamboyant all seem intriguing; 'plantsman's plants' is, I believe, the phrase, although this should not be used to devalue it.

Some purists decry *C. heracleifolia* and, I suppose, it does produce an inordinate abundance of leaf in relation to the amount of flower. Those leaves are certainly big, like a heracleum of course, and make a bold cover from which emerge upright stems with the flowers clustered in the leaf axils like so many dainty hyacinths,

the tip of each petal rolled back. Some forms are scented, but not all.

This is a difficult plant to place; it seems too coarse to be set at the front where the flowers can be seen and sniffed, yet further back between shrubs, perhaps, where its weed-suppressing qualities are of value, the flowers may hardly be noticed. Yes, the flowers. In 'Campanile' they are lavender blue, though a little deeper at the tips; in 'Davidiana' the colour is similar but the scent more impressive; in 'Wyevale' they are large and a rich blue, although small-flowered impostors sometimes lurk under this name. 'Jaggards' is a newcomer in deep blue – but I've not seen it. I raised a white-flowered form from wild collected seed – but somehow contrived to lose it.

Finally, come the undiscovered Americans. The particular feature of these American species is their slender growth and especially the endearingly rounded shape of their flowers. The four petals are fused for most of their length to create a dainty inflated ball, then they split and flare at the tip.

At the front of the border, in the delicate intermingling of plants which repay close inspection, the shorter of these can peep out appealingly. If you come across any of these names in catalogues or nurseries, place an order. Not all these names are valid, but they are sometimes in use: *C. addisonii, C. crispa, C. douglasii, C. fremontii, C. hirsutissima, C. occidentalis, C. pitcheri, C. scottii, C. viorna*.

'*Clematis scottii*,' says Claude Barr in his modern classic *Jewels of the Plains*, 'is a marvellously free-blooming leather flower with old maids bonnets, in soft violet-blue to deep purple, or a rare delectable pink... In the wild, a plant consists of one to several stems; under cultivation, well-developed specimens form a veritable mound as much as 20 inches wide, densely foliaged, with a corresponding bounty of blossoms.' Need I say more?

## Pinks and carnations

Carnations and pinks have been fixtures in gardens since before Parkinson confessed that so many were grown he had to treat them in separate chapters – carnations, gillyflowers and pinks plus sweet Johns and sweet Williams – and even so he could not develop the detail he felt they deserved.

Over the centuries there has been a shift in the balance of what is grown, especially out of doors, so now we grow few carnations in our borders and the breeding work of enthusiasts like John Galbally is appreciated mainly by those who grow for showing. The height of the plants, the need for staking and their less than robust constitution perhaps discourage us from growing them, but intimate and well-fed borders could surely be enhanced by a few border carnations, with support being provided as much by bushy neighbours as by brushwood.

Exquisite forms are still being raised and introduced. The white-flowered 'Irene Della-Torre', with its rich red picotee fading to blush pink, seems more floriferous and a little more robust than most, 'Ruth White' is similar but in a deep reddish purple set on white, while the lovely 'Eileen Neal' is deep pink red streaked in misty mauve.

In border pinks, we seem to be moving in two directions at once. Developments in breeding for the cut flower trade have led to highly productive, sometimes slightly larger-flowered varieties being introduced, particularly those in the 'Devon' series. Breeders have also responded to public demand, from shoppers as well as gardeners, by selecting varieties with a good scent and even the deep red 'Devon General' is scented. 'Devon Dove' is a rather frilly white with clove scent,

'Devon Glow' is a magenta with a strong scent and 'Devon Maid' a semi-double white with a magenta centre and a clove perfume.

At the same time the resurrection of old pinks is much in the air, but to some extent those who treat this with an undue degree of seriousness are hoaxing themselves. It seems unlikely that any pinks from hundreds of years ago, even perhaps from the last century, are still with us through direct descent by vegetative propagation and that similar seedlings are now doing duty for them. Is the heavily scented 'Charles Musgrave' which we grow today the same green-eyed white with slightly fringed petals that was grown in 1725? 'Bats Double Red' from 1707 seems uncommonly vigorous for such an old plant; perhaps this too is a more modern seedling. The old descriptions and stylized illustrations are rarely sufficiently precise for us to be truly certain.

In parallel with this looking back has come a more positive looking forward, for there has also been some selection of new varieties in the old style, especially in laced pinks. So following the clove-scented 'Gran's Favourite' in white with a pink lacing from 1966, a plant whose name is doubtless intended to suggest a much earlier origin, there was the single-flowered 'Constance Finnis' in 1969. This single white with a pink lacing looks superb with *Campanula patula* and *Penstemon* 'Stapleford Gem'. More recently we've had John Galbally's 'Becky Robinson' in pink with dark eye and lacing, 'Lincolnshire Poacher', a single lavender pink with a maroon eye, and 'Margery's Choice' in almost white with ruby-magenta lacing and eye.

At the other extreme are those which do not fit into the usual pattern at all. Very few bold single-flowered varieties are widely accepted but 'Kesteven Kirkstead' surely has a sufficiently captivating combination: pure white flowers with a bold, crimson eye. It must be unique. Then there are those which are connected with recent introductions from the wild. Since *Dianthus superbus* came in again from Japan as one of the collections from an Alpine Garden Society expedition its scented frilly flowers have been seen more and more, but although it knits well into front-of-border jumbles, it has a short life in the garden.

'Red Feather' and 'White Feather', singles which seem to be selections from *D. superbus*, are sold in Holland as annuals but seem to last fairly well as perennials, particularly in well-drained conditions. But an upright and upstanding habit is not one of their strengths so, like so many in this section, they need support from slightly more robust neighbours. Planted close they can grow through, rather than alongside.

In the same vein, a close-to-wild plant, is *Dianthus turkestanicus* 'Patricia Bell', a double-flowered form of the sweetly scented species selected from a batch of seedlings by John Fielding, with soft pink flowers fading almost to white; a very fashionable shade.

Dianthus in general seem overdue for a revival. Richard Bird's carefully researched new book *Border Pinks* will perhaps help them along, and the developing trend for thoughtfully managed, deeply interwoven borders of single plants which enmesh intimately in a constant evolution of flowers and foliage should suit them well.

### Peonies

It seems to have been a sad decade for peonies. Well, perhaps that is too pessimistic a summary, but those legions of old varieties now make but little trade through nurseries compared to their heyday and as I write Kelways Nursery, the name with which peonies are so solidly associated, has recently been rescued from financial difficulties for the third time in almost as many years.

With so many plants there has been a drift of interest from the highly bred varieties to the purer species. Twenty years ago the late autumn chrysanthemum *Dendranthema nankingense* would rarely have been noticed; at best it would have attracted a moment's attention before being quickly passed over. Then, when it first appeared at a late autumn RHS show a few years ago, gardeners fell on it with enthusiasm as a wonderful new autumn perennial at the same time as the highly bred florist's hybrids provoked increasing disdain.

The highly bred peonies have suffered a great deal less scorn; they seem simply to have slid quietly out of our consciousness. 'The herbaceous peonies stand straight and tall,' says Eleanor Perényi in *Green Thoughts*, 'don't hide their heads and are magnificent for cutting. They aren't temperamental, deciding, for inscrutable reasons, to withhold their bloom for a year. They are almost immortal, even when hopelessly neglected in the backyards of old farms. They can stand temperatures below zero.'

So why is not every garden bursting with them? She goes on: 'All peonies suffer when a heavy rain hits them.' That must be part of it, that and the price. Many are more expensive even than named hellebores, but there is not, of course, the opportunity for second-string seedlings to fire up new gardeners to their appreciation and so lead them on to the best. Peonies are slow to bulk up sufficiently for propagation and at a season when there is so very much which excites our interest, for most people their flamboyance does not quite make up for the fleeting nature of the display and the cost.

But they do have an opulence which would turn to garishness were the colours less refined. True, the Imperials with their bold boss of partially petaloid anthers can be very bright when, as in 'Emperor of India', introduced by Kelways in 1901 and still with us, the golden anthers are set against bright red petals. But generally the peony is a harmonious flower.

However, in recent years it is the species which have captured our attention. The current enthusiasm for species plants, for obscure wild forms of the highly bred and the familiar, seems to make up for the fact that the species are often even more fleeting in flower.

Two have particularly grown in popularity and are now raised in ever-increasing quantities. *Paeonia mlokosewitschii* is one of those plants whose qualities justify its being fashionable. Simple, single, pale lemon flowers sit above slightly greyish foliage in quiet confidence, followed by sparkling seed pods which split to reveal smoky blue fertile seeds mixed with infertile seeds which are bright carmine. 'Mollie-the-witch' it has become known as, in response to what Farrer called 'this pleasant little assortment of syllables', but, like the more highly developed sorts, its slow progress towards flowering has restricted its spread.

The other species, one which has been admired by alpine growers for many's the year and even grown in pots in the alpine house, is the Balearic endemic *P. cambessedesii*. 'A gem for picksome people,' says Graham Thomas, which seems to carry an edge of condemnation although I'm sure this was not intended. Shining red stems hold silvered green leaves backed in the same red, and as the leaf edges roll, the undersides glint in the sun. The flowers are pink and come earlier than most, in April even, on short plants.

This species is fussier than most. It demands full sun and a crisply drained soil and its rather delicate early growth needs protection from frost. But those fragile pink bowls set against silvered foliage are simply beautiful.

There is one more for which demand always exceeds supply, *P. tenuifolia*. This species too has good foliage, freshly green and finely dissected,

and the shining crimson flowers almost float on the leaves. There is a pink, which is less effective, and a double red which makes up in the longevity of its flowers what it lacks in simple elegance. There are rumours of a form with broader leaf divisions – which seems pointless.

The highly developed *P. lactiflora* hybrids are slow to bulk up enough for propagation, and this restricts their availability and the possibilities for general nurseries to carry a good range. With the species, it's the unsuccessful pollination, slow germination and slow growth to maturity which cause the problems.

In peony species it is often the case that single plants will set seed only if they are pollinated by another, genetically different, plant. So however flourishing your one precious plant, you may not be blessed with seed. Then there are germination problems. Peony seed starts to germinate after one cold spell but produces only a root; after a warm spell and another cold period the shoot appears. Seed pots discarded after one winter when nothing seems to be happening may conceal a root which has started to grow. The long delay before growth begins in earnest also allows plenty of opportunity for slugs to cut down your hopes.

Having said that, I was astonished to discover shoots emerging from sowings of *P. broteroi* from two different sources emerging just a few months after sowing. There seem few rules which cannot be broken.

### And finally for summer

Summer fills the garden with so many good things that it could just as easily fill this book – in order to allow any space at all for the other seasons, I must examine but a selection of summer plants. Having made one choice for the main entries, then a second choice of clematis, dianthus and peonies in this chapter, I simply cannot pass by so many others entirely. But I know my borders

The dusky leaved 'Solfaterre' is the most distinctive of all crocosmias

suffer from just this fault – I always want to cram a few more things in.

Acanthus are often praised for their fulsome foliage but I suspect this is partly because not all gardeners grow them sufficiently well to produce the flowers. More or less by chance I found that *A. spinosus* 'Spinosissimus' thrives in the combination of heavy soil and full sun in an open position, not only bulking up well into great folds of glossy foliage, but throwing a battery of flower spikes in August and September with a few even later. But the one I covet is *A. dioscoridis* var. *perringii* – I say covet, in fact I already have the plant, but not the flowers. This exquisite pink-flowered form is less tough than *A. spinosus* so I gave it better drainage; but so far it's the usual problem, lovely soft greyish foliage, increasing every year, but no flowers.

Crocosmias are enjoying a most welcome revival. Around 125 different varieties have been recorded over the years but by 1976 only twenty-three were available from nurseries; now the numbers are up again and forty-six are listed. This change has come about in two ways.

First old varieties are being rediscovered, especially in Ireland, where the climate suits them so well. These are now being made available by nurseries – sometimes their old name has remained with them, but sometimes the original name has been lost and as old descriptions are so vague, it can prove impossible to match them up with a true name.

Second, new varieties are being bred. Bressingham have introduced nine, most of which they have raised themselves, and Treasures of Tenbury introduced about the same number though with rather less fanfare, so little in fact that they made but a modest impact; I found them less tough than most.

Out of all these, old and new, I pick five: 'Jenny Bloom' in soft creamy orange; the vibrant and vigorous 'Lucifer' in fiery red; 'Mrs Morrison', recently discovered in Ireland, with bold, open, tomato red flowers; 'Solfaterre', with apricot yellow flowers above smoky bronze leaves; 'Star of the East' which dates from about 1912, with large-flowered, outward-facing orange flowers.

Some plants simply don't excite me, hemerocallis for example. Many American breeders — and almost all the new introductions come from the States — seem to have been infected by the same malady to which the breeders of bedding plants are so prone, an inclination to ever smaller plants with ever larger flowers. The result is a range of plants which may be impressive in shows but not in the garden.

So it's back to the old days and plants which throw such a constant profusion of small flowers that the fact that any one only lasts a day matters but little. Of those given an Award of Garden Merit by the RHS — and many of the most recent American introductions have been trialled over the years — all were introduced at least twenty years ago and most in the 1950s. 'Stella de Oro' is the most recent.

So cannot breeders look to the habit of varieties like 'Corky' and 'Golden Chimes' and introduce new plants in the same style but in a wider range of colours?

No one has spent any time breeding perennial lathyrus, as far as I know, although I'm tempted to say that they are of more use in modern gardens than hemerocallis. By growing them through shrubs and shrub roses delightful associations of colours can be created, and when trained through forsythias, weigela, lilacs or other shrubs with a fleeting period of interest a second season of flower can be created in the same spot. The secret is not to train them up fences or wigwams of stakes as was once the habit, but to plant them under shrubs or small trees and let them scramble through to emerge, flowering, into the light.

The bricky red *L. rotundifolius* is perfect lacing its trailing spikes among the soft pink flowers of 'Heritage' roses, and where the support is sufficiently solid and perhaps a multistemmed tree can be the host, both lathyrus and clematis can combine.

The so-called perennial sweet pea, *L. latifolius* is the most widely grown, though this misleading name hints at scent. There are magenta pink, paler pink, blushed white and pure white forms which are sometimes given glorification by names like 'Pink Pearl' or 'White Pearl', but it's hard to say whether these names are justified. John Metcalf's 'Blushing Bride' is a delicately coloured exception.

The perennial lobelias are something of an enigma. Once grown as summer bedders or waterside plants and brought into the protection of a cold frame for the winter, they are now on the march into mixed and herbaceous borders and also into the cut flower markets.

For years seed strains like 'Queen Victoria' were grown with a few named forms to be propagated vegetatively. Then Dr Wray Bowden in Ontario began crossing forms of *L. cardinalis* and *L. syphilitica* and the result was a large series of new introductions, all tetraploids, with stout stems, large flowers, strong rosettes and fibrous roots. They are also hardy, this hardiness being derived from a form of *L. cardinalis* from southern Ontario and one of *L. syphilitica* from the same area. His aims have been realized in his many varieties which are tough, come in a wide range of strong colours with long racemes of flowers. Many also have at least a hint of purple tinting in the leaves.

Relatively few of his varieties are generally available, in spite of the fact that he has named at least eighteen, but 'Cherry Ripe', in cherry red ageing to rose and 'Will Scarlett' in brilliant red are seen. His work has been carried on by Johannes Kriegs of Ernst Benary Seed Growers in Germany, and this has led first to the introduction of the extraordinarily vigorous $F_1$ hybrid 'Compliment Scarlet', intended mainly as a cut flower but exceptional in the garden. This has been followed by companion varieties in deep blue and deep red, all with a non-branching habit. Now there is also 'Fan Cinnabar Rose', 'Fan Deep Red', 'Fan Orchid Rose' and the very impressive dark-leaved 'Fan Scarlet', which branch more strongly from the base.

These are all good perennials if protected from slug damage in winter, and I found that plants of 'Compliment Scarlet' produced three or four spikes in their first summer if raised in the same way as bedding lobelia then pricked out into individual pots. In their second summer some plants produced eleven spikes.

Breeding in lupins had been quiet for many years as the old Russell lupins, raised from cuttings, slipped away and were replaced by increasingly variable seed-raised varieties. Then Woodfield Brothers of Stratford-upon-Avon began to revive them, with the particular aim of ensuring that the flowers at the base of the spike lasted until the flowers at the tip opened. They have raised some impressive forms, in pure shades

and bicolours, and their annual stand at Chelsea shows them off well. 'Troop the Colour' in deep red, the shining yellow 'Moonraker' and the pure white 'Deborah Woodfield' are especially fine but not easy to come by.

Finally another group where breeding work has made great strides in recent years: monardas. 'Cambridge Scarlet', 'Loddon Crown' and 'Croftway Pink' were the old dependables in red, purple and pink, the distribution of seedlings not enhancing their reputation. Then Piet Oudolf in Holland began to make selections with mildew-resistance in mind; his varieties fall into two series. Plants in the Indian Tribes series are tall and show some mildew-resistance, those in the Signs of the Zodiac series are shorter and with excellent mildew-resistance.

All are worth trying, although there has been some ridiculous confusion over the names of some of those in the Signs of the Zodiac series, with 'Pisces' being translated into 'Fishes', and I've even seen it labelled 'The Fish'. We also have 'Bowman' for 'Sagittarius' and 'Scorpion' for 'Scorpio'. The British and their flair for languages... Who else would see the necessity of translating Pisces to fish? It makes translating *Achillea* 'Hoffnung' to both 'Hope' and 'Great Expectations' seem almost sensible by comparison.

*Opposite*
*Dahlia* 'Fascination', *Tricyrtis hirta*, *Schizostylis coccinea* 'Jennifer', *Dendranthema* 'Anastasia', *Aster* 'Monte Cassino', *Phytolacca americana*.

LATECOMERS

# LATECOMERS

The poets, you know, got it all wrong. Keats and the others who took nature for their theme are at their most seductive singing lyrically about spring and summer. They're also very persuasive when they turn to autumn, but suddenly it's all mists, owls, departed swallows and dead roses. Shelley even titles one poem about autumn simply 'A Dirge', which about sums up their attitude, and Tennyson had obviously had quite enough when he wrote:

*The air is damp and hushed and close,*
*As a sick man's room when he taketh*
*repose*
*An hour before death...*

As for Thomas Hood:

*No warmth, no cheerfulness, no healthful*
*ease*
*No comfortable feel in any member –*
*No shade, no shine, no butterflies, no bees,*
*No fruits, no flowers, no leaves, no birds –*
*November!*

Sadly, it was around that time of year in 1844 when autumn finally got the better of him and poor Mr Hood, who I must say had plenty of lighter moments, took to his bed – and never got out of it until he was carried down the stairs in a coffin six months later, aged forty-six.

Autumn may have done for Thomas Hood, but neither he nor the rest of them can have popped their head around the gate of a good garden on a sunny autumn afternoon – or a misty one for that matter. For although even now some gardeners despise autumn for its windfalls, its mildewed Michaelmas daisies and its soggy bedding, there's as much excitement in an autumn garden as there is in summer and a great deal less gaudiness.

And what treasure buried in catalogues among more familiar favourites. Anemones in red, pink and white, single and double; asters, from hummocks to head-high, with, and especially without, mildew; clouds of white and pink boltonias; tropical cannas; chrysanthemums in every shape and hue; contented colchicum and crocus, spreading happily; cyclamen, with marbled foliage to follow; swaying purple eupatorium; helenium and helianthus, the yellow autumn daisies; arching pink lespedeza; blue liriope; miscanthus, in misty clumps; salvias in penetrating blues and fruity reds and purples; sedums for butterflies; tricyrtis, the intricate toad lily – and more.

I can give but a selection of autumn flowers; some will be familiar, some names may spark a faint recollection, some will be unknown. But with the foliage of evergreen shrubs and perennials, with so many autumn-colouring plants, berried and fruiting shrubs, lingering annuals and the rest, there is no reason to spend the autumn drooling over catalogues of spring bulbs and summer annuals, blind to the riches of the undiscovered season which could be there, outside your own window.

Like spring, autumn is an unpredictable season; the onset of autumn colour sometimes seems almost mystical in its timing, controlled by some unfathomable inner rhythm. The profligacy of berrying plants often depends on the weather many months before, and it can be hard to recall quite when those icy winds kept the bees in the hives or when that frost fell.

Frost in autumn is a mixed blessing, for while dahlias almost instantly succumb to its bite, it sets the foliage turning. An early snap to dispatch the weaklings and start the colouring process can be quite helpful, but it seems sad that late-sown half-hardies, which can add so much, should have their term cut short.

But the truth is, we just don't need them. The light of autumn, at times sharp, misty, or soft and flattering, makes so much of every autumn treasure. Allen Lacy, gardening columnist of the *New York Times*, says it all: 'Autumn is the season of flame and fire and incandescence, of scarlet and crimson and bright gold. But the palette of autumn also embraces more subtle tones and shades. To name the colours of autumn is to speak of lavender and lilac and buff rose, and finally to give up the effort, since at this season of the year there are so many colours with no names in any language.'

# Chrysanthemums

For once, I am delighted to be able to approve of traditional expectations: what, after all, is the point of chrysanthemums which flower in August? The garden chrysanthemum is a flower of the autumn and it is surely perverse in the extreme to turn it into a summer flower. Think about it; a gardener with little imagination can ensure that the borders are colourful in August simply by following the instructions in a magazine. So why try to breed chrysanthemums to flower earlier and earlier when it is in October and November that we need them so much? More colours in more forms for late-season borders would be most welcome.

While breeders are attempting to create varieties which flower earlier and earlier, their blinkered imagination struggles with the concept of novelty; their attention seems incapable of distraction from ideas of ever dwarfer habit and increasingly bizarre colour combinations. Dwarfness, that I can just tolerate, although breeders should perhaps realize that this is not the only way to ensure plants are self-supporting. So 'Yellow Flare', given the prestigious FCC by the RHS in 1991, scores for just one feature, its neat yellow pompoms. But they are at their best in August, the plants are just 15in high, and the flowers age to a ghastly dirty pink. Horrid. Horrid. The taller 'Wendy' in rusty orange is altogether more acceptable and a wonderful autumnal colour, but in August?

## Picking through the rubble

There was a time when the genus *Chrysanthemum* comprised about 200 species in a variety of forms. In one of the best known of recent demolition jobs, botanists have stripped to their vests and broken up the genus so that all that remains is the hard core of admittedly very lovely Mediterranean annuals.

The remaining rubble seems to comprise mainly *Arctanthemum, Argyranthemum, Chrysanthemum, Chrysanthemopsis, Leucanthemella, Leucanthemopsis, Leucanthemum, Nipponanthemum, Pyrethropsis, Pyrethrum, Tanacetum,* plus of course *Dendranthema** – and in this new genus we find the hardy garden chrysanths, those related to the florist's cut flower and exhibitors' chrysanths. In a way this too, like the old genus *Chrysanthemum* in microcosm, is something of a mixed development. There are the exhibitors' chrysanths in their various forms, some grotesque, some relatively attractive. There are pot 'mums, dwarfed by chemicals. The cut flower chrysanths are mainly sprays grown under glass but also include outdoor sprays, some of which are also grown in gardens. In all, the National Chrysanthemum Society (a change of name to the National Dendranthema Society seems as likely as the Conservative Party deciding to call itself the Bleed the Rich Party) has classified the genus in thirty sections, almost all of which we can happily consign to the skip, retaining only those in two sections, 28-Pompoms and 29-Sprays, for inspection.

Having done that we can further enrage the

---

* Now, it seems, there are plans to reunite them all in *Chrysanthemum*... These botanists!

NCS by all but abandoning what small part of their system applies to us gardeners; after all, the NCS makes no effort to cater for us. In spite of the fact that Section 29 has been conveniently subdivided into six subsections, two of the most important groups to the gardener, Koreans and Rubellums, have not been given specific allocations. So let us follow gardeners' convention and retain three groups: Japanese Poms, Rubellums and Koreans.

The origins of all these groups of garden chrysanthemums are confused, to say the least. What is clear is that without the various importations of cultivated and wild Chinese and Japanese sorts, none of the European developments would have been possible. By the time the first plants reached Europe at the end of the seventeenth century (they did not survive) chrysanthemums had already been grown and hybridized in China for hundreds of years. There was a more successful introduction in 1789 and this rapidly developed into a stream of such arrivals. Breeding began and by 1842 there were over 100 varieties grown in Britain.

These were all based on what we now call *Dendranthema indicum*, with small, single, yellow flowers, although a purple-flowered form, 'Old Purple', was one of the first introductions. In 1846 Robert Fortune brought back the Chusan chrysanths, miniature precursors of pompom types, then in 1861 he introduced from Japan forms with reflexed petals, although these were frowned upon at first. Most developments occurred first in the East, although Europeans were responsible for developing varieties which flowered early enough to be grown outside.

### Japanese poms

Many of the Garden Chrysanthemums are generally included in NCS Section 28, and these include the well-known pompoms like 'Mei-Kyo'

and its sports plus one or two others. 'Mei-Kyo', which means 'Treasure of Kyoto', is a very pretty little soft pink pompom. In the late 1950s Will Ingwersen was offered cuttings of this plant by a correspondent in Japan. They arrived in a matchbox, were rooted and introduced in his 1960/61 catalogue.

Ten years later this variety sported in the garden of one of Ingwersen's customers. He passed stock back to the nursery, who introduced it as 'Mei-Kyo Bronze Elegance' in 1973. Then in 1988 a white sport occurred on 'Mei-Kyo' in an Essex garden and this was named 'Purleigh White', a pun on the name of the village in which it occurred. This was introduced in 1990, and then in 1991 Ingwersen's showed a yellow sport of 'Bronze Elegance' which had occurred in the garden of Rose Clay at Nantyderry near Abergavenny in Wales in 1989. This was named 'Nantyderry Sunshine'.

Tough, reliable and easy to propagate, the leaves of all these forms are as neat as the flowers. They open in September and October and have two cousins whose origins are less clear. 'Dr Tom Parr' is a soft pinky brown while 'Anastasia' is always said to be similar to 'Mei-Kyo' but a little earlier, a little paler and a little heavier in its bearing. But 'Anastasia' and 'Mei-Kyo' are so similar that they've become hopelessly confused. 'Anastasia' was mentioned by Sitwell before the war, but even then he remarked on the difficulty of distinguishing it from 'Little Bob' and 'Firefly', both now lost. It has an unpleasant variegated form; you would not buy it, I am convinced, and if you are given it you can only hope it will soon revert.

Some light has been shed on this confusion by a Hardy Plant Society member, John Gregory, who did what any sensible gardener would when confronted with such a muddle and grew 'Mei-Kyo', 'Anastasia' and 'Anastasia Variegated' side

Three of the best hardy chrysanthemums: 'Bronze Elegance' (*top left*), 'Mary Stoker' (*top right*) and 'Innocence'

by side. 'Mei-Kyo' and 'Anastasia' proved identical in every respect, even down to the number of petals in the flower, while the plants known as 'Anastasia Variegated' were at least 12in shorter, with very slightly larger flowers, with far more petals and more of a reddish tint – and of course they were variegated. This seems to prove that the variegated form has no connection with the plant from which it is said to have sprung.

Be all that as it may, this group is indispensable in the autumn garden. Their self-supporting growth needs the minimum of management; their foliage is neat and dark, not soft and flabby like

that of so many chrysanths; the flowers open late, not usually before about mid September, and last well; the neatness of the flowers combines perfectly with their total lack of either flaunting flamboyance or that insufferable arrogance of the chrysanths we are sometimes encouraged to grow in gardens.

Some grow 'Mei-Kyo', in soft pink with its peeping yellow eye, in a row for cutting, and if you have the space I would have a plant or two of each of these specifically for the purpose. I grow it behind the little *Aster novae-belgii* 'Professor Anton Kippenberg', spiky-flowered, the blue rays hiding the eye and its dark, bushy growth usefully covering the one fault of 'Mei-Kyo' and its relations, a tendency for the basal leaves to wither. I am tempted to put *Carex buchananii* in between and cut it back hard in May or June to keep its growth a little more modest by autumn than would follow a March cropping.

Pure white *Colchicum speciosum* would squeeze through and open its snowy chalices at a suitably harmonious moment, and our old friend *Aster lateriflorus* 'Horizontalis' could sit alongside 'Mei-Kyo' and ease the transition to the next part of the border.

I would be tempted to plant the rediscovered *Aster novi-belgii* 'Calliope' with its single blue flowers and almost black stems behind, but it is just a little too tall and pinching in June would deprive us of those fat, luxuriant stems. So instead, perhaps, *Salvia uliginosa*, which can stand a little pinching and if it falls forward will be steadily supported by 'Mei-Kyo'.

Another chrysanth for this planting would be that ancient cottage favourite, 'Emperor of China'. Sometimes known simply as 'Old Cottage Pink', this is one of the very last to flower, the frost having edged the leaves with red by the time the flowers open. Any plant which waits in the wings until November and then stands firm against wind and cold deserves our applause. Fortunately its flowers are in tune with the season, silvery rose pink and almost crimson in the centre, and although its lowest leaves may wither, by November I think we can forgive this fault.

Just thinking about these warm, comforting, prepare-to-curl-up-in-front-of-a-log-fire plantings sparks ideas for other companions. *Schizostylis coccinea*, either the pale and vigorous 'Jennifer' or the rich and shimmering 'Hilary Gould'. *Liriope spicata* in blue or white, and another

unjustly ignored wonder of autumn, *Elsholtzia stauntonii*.

### Rubellums

The Rubellum chrysanthemums grown in gardens today are generally reckoned to be descended from one plant, found growing in Happy Valley public gardens in Llandudno in 1929. There has been some speculation that this original plant was a hybrid between the long-flowering *D. zawadskii*, with pale pink flowers and finely divided foliage, and the invaluable *D. yezoense*, with late white daisies. This original plant is still available, under the name *D. zawadskii* var. *latilobum*, with pale pink flowers rather similar to those of 'Clara Curtis'. This plant, and the group as a whole, is characterized by a tendency to produce runners rather than remain in a tight clump.

Once introduced to Kew the original plant was exhibited in 1935 and given an Award of Garden Merit in 1938, but by then Kew had passed seedlings to Gerald Perry at Perry's Hardy Plant Farm, for in 1937 he was able to introduce 'Anna

---

*Opposite*
The extraordinary *Delphinium* 'Alice Artindale', introduced in 1936 and seen here in John Fielding's London garden, is the only one of the old, fully double delphiniums which is still fairly easy to find in nurseries.

*Overleaf (left)*
ABOVE The glossy, mildew-resistant foliage of *Phlox maculata* 'Miss Lingard' is a most valuable feature, although the flowers in their cool white cones are sometimes tainted with a blushed eye.

BELOW Many of the older hemerocallis are the best for general border use. 'Golden Chimes' was introduced in 1954 and forty years later was still good enough to gain an Award of Garden Merit after trialling at Wisley alongside more modern garden varieties.

Perry' and 'Clara Curtis'. These were followed in 1939 by 'Elizabeth Cowell' but all seem very similar: 'Anna Perry' was described by Perry's as having 'deep rose-coloured flowers', 'Clara Curtis' was 'a delightful shade of deep rose-pink', while 'Elizabeth Cowell' was 'a delightful shade of bright rosy crimson'.

After the war Perry's introduced many more varieties and in 1952 they wrote: 'Each year at our selection trials we are faced with an embarrassing and bizarre array of wonderful blooms, so the task of selection becomes annually more bewildering and only the best that are improvements on existing cultivars are introduced.'

By the 1960s Rubellums were very widely grown. At about this time Perry's reprinted this extract from the *Times Survey of Gardening*: 'Still newer, and of great promise for the future, are the forms of *C. rubellum*. The Rubellums outshine even the various Korean strains, in hardiness, range of colour and freedom of growth. They seem to flower unchecked no matter how hard the autumn nip, and to grow them in a roadside garden is to invite an almost embar-

---

*Previous page (right)*
ABOVE 'Ellen Houston' is an invaluable plant for the autumn border, and in the dahlia trials at Wisley its small but richly-coloured flowers and bronze foliage are most welcome among all the show varieties.

BELOW Korean chrysanthemums are undergoing a welcome revival, and here in the autumn borders at Wisley the single-flowered 'Martha' interweaves with the rusty pompoms of 'Brown Eyes'.

*Opposite*
The de Caen anemones, developed from the Mediterranean *A. coronaria*, sometimes yield individual plants of great simplicity and beauty, like these white-eyed forms in John Fielding's garden. These can be selected and increased.

rassing degree of attention and enquiry from passers-by.'

But interest waned; perhaps this decline was fostered by the National Farmers' Union who urged the destruction of old clumps of chrysanthemums under the entirely misplaced conviction that they were hosts of white rust disease.

By the late 1980s only about half a dozen were available from nurseries, but the revival is now beginning. The Cambridge group of the NCCPG has collected together over twenty for the National Collection and these are the ones which seemed to me to be the most valuable, although all are worth growing.

'Alison Davis' is pale peachy red streaked with rose and then fades to rose streaked with white. The backs of the petals are slightly silvery, the foliage is edged with scarlet. Most are single but 'Duchess of Edinburgh' is double, although the yellow eye is revealed as the crimson flowers open wide; it's also the most floppy. 'Innocence', at its best at the end of October, is misty pink with stems tinted red right up into the flower head; it's dependable and flowers prolifically. 'Lady Clare' is similar but perhaps even more floriferous, and with the petals slightly streaked in silver leading to the old flowers becoming slightly blotched.

'Mary Stoker' is a fine pale buttercup yellow but develops pinkish tints as the flowers age; some like this mix of shades, others hate it. 'Nancy Perry' is a pale watery red but the flowers are large, at least 3in, although unfortunately the individual rays do not unfurl together so the result can be rather messy. 'Jessie Cooper' is similar but the flowers are smaller.

Frances Perry, who knew better than most, recommended the following varieties in 1957: 'Anne, Lady Brockett', 'Clara Curtis', 'Duchess of Edinburgh', 'Jessie Cooper', 'Lady in Pink', 'Mary Stoker', 'Paul Boissier', 'Pink Pearl',

'Queen of Scots', 'Red Ensign', 'Royal Command', 'White Wedgwood'. Most are represented in the National Collection and so there is the potential for us all to grow them.

Old varieties are still being rediscovered and the latest to reappear is 'Anne, Lady Brockett'. This was first grown around 1945 but seemed to be extinct until Joe Sharman of Monksilver Nursery noticed a chrysanthemum simply named 'Brockett' in a large French wholesale nurseryman's catalogue. He made the eight-hour drive from Cambridge to Paris to acquire it, and now this large-flowered, dusky pink form with apricot buds is back in cultivation in Britain and getting around more widely.

### Koreans

The first Korean chrysanthemums were raised in Bristol in 1937 and it is worth reiterating that this is Bristol, Connecticut, not Bristol, England; perhaps this distinction, sometimes overlooked in Britain, explains their hardiness.

The Koreans are said to have arisen from crossing the original Rubellum with existing florist's varieties, but the timing seems very tight. Half a dozen were listed by Perry just a few years later and by 1939 Blooms listed twenty-six. A long list was available in the 1970s, then there was a slump, then a resurgence and now a wide range is again to be found. A huge number were named over the years; Home Meadows Nursery now lists over forty, but none of Alexander Cumming's originals which were named after the planets. Koreans are characterized by their hardiness (Connecticut is mainly in zones 6 and 7, Britain mainly in zone 8), their generally late flowering, the variety of flower forms and colours and the fact that they have a good garden habit; they look like border perennials rather than stiff and starchy like florist's chrysanths.

The first I grew was 'Wedding Day', one of the later introductions and at 4ft one of the taller. Its white daisies have the special attraction of green rather than the usual yellow eyes and this gives it added distinction. Mrs Fish treated this plant like a herbaceous clematis; planted underneath a *Phlomis fruticosa* it grew up through the middle, the branches of the phlomis providing a strong support and its grey-green leaves a suitably soft background.

It has to be said that as a group the Koreans have become increasingly difficult to define, as almost any fairly hardy chrysanth seems to be included under that banner. They vary in height from 18in to 4ft, the flowers may be pom-poms, singles, spoons or anemone-centred as well as the usual doubles, and while some flower as late as November, others like the peachy pink 'Aline', with its spoon-shaped rays, flower too early to meet entirely with my approval.

Making a choice is not easy, but in addition to 'Wedding Day' I would select 'Brown Eyes', in a rich two-tone browny orange, for its invaluable autumnal colouring, 'Martha' in a suitably autumnal red and the recently rediscovered 'Tapestry Rose', a semi-double in deep pink with the green eye of 'Wedding Day'.

More recently the situation has become yet more muddled. Chrysanthemum specialist Rileys of Alfreton in Derbyshire have developed the Pennine series and over the years have introduced a steady stream of relatively hardy varieties, all with the Pennine prefix and in a very wide range of colours and forms. More recently we have seen the appearance of the Bird varieties, which now include a number not named after birds at all. These are relatively dwarf and very prolific and there are a few, like the even shorter 'Golden Plover' at 6–8in, which is confusingly placed in the Mini series.

In addition, there are all those garden spray chrysanths most of which are less hardy but which include some of the prettiest of all, notwithstanding their early season: 'Dee Gem' is a dark-centred, pale pink double, 'Wendy' is a suitable autumnal orange double, while 'Wessex Lady' is a good strong-coloured single pink. But I must emphasize that these sprays should be examined carefully for good garden characteristics. Some are stiff and ungainly, in some the flowers are too congested and this spoils the effect, others fade inharmoniously. There is no substitute for looking and judging.

A little thought is appropriate when considering how to perpetuate these varieties. Traditionalists tend to lift all their chrysanths after flowering and store them in boxes in a frost-free place for the winter. They are then potted in spring, cuttings may be taken, and they are replanted after the worst of the weather has passed.

On very heavy soils, in very wet areas and in very cold climates all this will be necessary. But on lighter soils and in drier and warmer conditions this is less vital. My view is that border chrysanths should be as tough as my other border perennials, but unfortunately, in my various gardens over the years, different varieties have proved the toughest.

### Simple pleasures
In addition to all these highly bred varieties there are relatively untainted wild species, simple, unsophisticated beings unmolested by the hybridizer's tweezers. When I first saw *Dendranthema yezoense* in full flower in November yet on stems only 12in high over smooth, fleshy leaves, I could hardly believe such a lovely, unaffected plant could possibly be at its best at such a late season and still remain so small. Of course now, twenty years later, I realize that almost nothing is impossible among plants.

If the flowers of *D. yezoense* were quite as white as a Shasta daisy they would gleam too glaringly in the soft autumn light. But they are of a shade which might be called 'dirty' in high summer but which in autumn is hardly noticed. This species is a little more particular than all those hybrids. It needs well-drained soil to go with its full sun, and I find it best divided and replanted every two years otherwise it seems to fade away.

A much more recent introduction is the even later-flowering *D. nankingense*, with little yellow daisies which if they opened in July or August we would instantly dismiss, but in October and November and even at Christmas... Reaching about 2ft, the effect is of a hazy yellow cloud over bright green foliage, and being tough enough to withstand frost yet not at all overpowering it fits well with many autumnal shades.

# Dahlias

I am, perhaps, overstretching the concept of the hardy perennial more than will be acceptable to many of my readers by including a chapter on dahlias. Not only are the vast majority of dahlias not at all hardy, but many enthusiasts for hardy perennials would give up gardening rather than have dahlias in their borders. They regard visiting a garden for the first time and discovering dahlias rather like going to a dinner party and being served cod and chips with a bottle of light ale; not quite what they had in mind when they were getting dressed up to go out. And not at all what would be served at *their* table. Dahlias, they're just so colourful, you see, and the last thing the collectors of hostas and obscure epimediums require is colour.

This is not a view to which I subscribe (in my house, fish and chips is a treat), otherwise perhaps I would have included a whole chapter on *Boltonia* in place of this one on dahlias (you'll find them mentioned in the final chapter).

## Civilized society

I cannot but agree, of course, that many dahlias are indeed gross and have no place in my borders or my book. But perhaps I am thereby excluded from certain strands of civilized society, for I was astonished to find Sacheverell Sitwell, writing in 1939, after a thoughtful discourse on florist's primroses and pinks, coming approvingly to dahlias: 'The Show Dahlias are the huge, quilled flowers with wonderful colours and symmetrical, as it were, petal cells. Their popularity faded away fifty or sixty years ago. And yet, by some fortunate chance, they still survive, though regarded with suspicion by those persons who are ignorant of beauty. [But] these Show Dahlias which, alone, perhaps, among flowers in our day, reveal to us the standards to which the old florists could attain, form a whole taste to themselves and become one of the delights of autumn.'

But no, they always were a funny lot and a taste for monster dahlias was probably singularly unremarkable by Sitwell standards. And there were many who did not consider their society at all civilized.

Mr Bowles was a discriminating gardener who had a problem with dahlias: 'I should like to be strong-minded enough to dislike Dahlias and to shut the garden gate on one and all of them as a punishment to them and the raisers who have produced some of the horrors of modern nightmares. But as it would not make much stir in the world of Dahlias however tightly I barred it, and as I have a great affection for single dahlias and all the true species I have been able to get, the family is still admitted.'

He discriminates, but few members of the family were allowed through the gate into the garden at Myddelton House. 'I cannot believe I shall ever be converted to a taste for Collarette dahlias... Paeony-flowered dahlias that get larger and more violently glaring each season... I am not fond of Cactus Dahlias.' He then goes on to describe some singles and decoratives that he finds acceptable, mainly for cutting, and then we are back full circle to the one dahlia regularly admitted to the border by enthusiasts for hardy plants: 'My favourite of all Dahlias is a true species, *D. Merckii...*'

Discrimination, that is the key; 'there is no justification for sweeping them aside with a dismissive gesture as vulgar or clumsy,' says Christopher Lloyd, but he too, is, quite rightly, choosy. Mr Sitwell enjoys the florist's monsters, Mr Bowles enjoys the delicate species, Mr Lloyd fancies the 'Bishop of Llandaff'. Only Mr Sitwell is out of tune with today's music. How, then, am I to choose which are to be saved and which left out to rot?

If my garden were open to the public, I would be tempted to describe it in the guidebooks as a plantsman's garden concentrating on hardy perennials. Then, immediately inside the gate, I'd have the visitors turn a corner and suddenly come across a dahlia garden filled with the boldest and biggest and most outrageous varieties in a dazzling fury of colours. I would have that quotation from Sitwell engraved upon a stone for them to read. And as they staggered away, rubbing their eyes and puzzling over the text in the *Yellow Book*, around the next corner there would be a tea room and a green lawn surrounded by a dark yew hedge where they could quietly recuperate.

For the purposes of this inquiry I have but one qualification which varieties must possess in order to be admitted. It is not the one cited by the American writer Henry Mitchell: 'Dahlia fanciers, who like all horticultural fanatics, tend to be somewhat lopsided in their enthusiasms, profess to see great delicacy of shape among dahlia flowers, and to hear them talk you'd think these great, flamboyant daisies had every elegance, every grace. Let us admit it once and be done with it: the dahlia somewhat lacks the charm of the lily of the valley, the dramatic tension of the iris, the fragrance of the nasturtium, and so on. What it does offer is a brazen contentment with its flaunting color.'

For as a result of this 'flaunting color' he consigns dahlias to 'a sunny field among corn and pumpkins'; an idea worth trying if you have the space but completely ignoring their worth in more controlled and more valuable situations. Actually, Henry Mitchell has some funny ideas about dahlias and how to grow them. 'Few labors are more pleasant than buying five-foot stakes (once one recovers from the criminal price charged for them) and painting them green in the basement on damp, cold spring nights.' Personally, I'd prefer to spread newspaper over the living-room carpet and paint them green in front of a log fire with *The Rite of Spring* going at full blast and a large glass of Jameson's Irish whiskey to hand. Even then many labours would be more pleasant, including sticking to the log fire, the Stravinsky and the Jameson's and forgetting about stakes altogether. Or buying them ready painted at no extra charge.

Be that as it may, I choose to consider only dahlias which, hardy or not, fit into the mixed and perennial borders which are such fundamental features of gardens in temperate climates. By their colour and their demeanour they must belong – otherwise they must be shut out.

### The border dahlia

It is surprising how rarely dahlias are grown in mixed or herbaceous border plantings. True, there is more work involved than with the majority of the hardy perennials, but do we enjoy our gardens simply for what they look like from a deck chair or also for the pleasure in creating them?

These days, when so many more frost-tender plants are grown for the summer garden, the time spent dealing with dahlias seems less of an extravagance. But why, given the necessity for regular attention, should we bother? There are four answers: the unsurpassed array of flower colours, the variety of flower forms, the long flowering season and the basic one of the prodigious supply of flowers. Apart from the care they require, there is just one argument against them: inelegance.

Attempting to assert that dahlia plants present an attractive feature in their very form and habit would be pointless, although some are a great deal better than others. The main characteristic to look for is long stems on the flowers, as this ensures that they are held well above the foliage rather than squashed in among it so that the whole plant resembles a green pudding dotted with Smarties. 'Ellen Houston' is poor in this respect, 'Fascination' is much better.

Another feature to consider is the size of the flowers in relation to the size of the plant as a whole. A large plant is better able to take large flowers, as not only is it more likely to have the strength of stem to support an 8in bloom but the proportion of flower to foliage is more satisfying; small flowers, as long as there are plenty of them, also look good on a large plant while varieties producing large flowers on small plants should be passed over. Small flowers on small plants usually look well, although some varieties can carry too many flowers, and begin to look like those horrid African marigolds so beloved of city parks departments.

Branching habit is worth considering, although this can be governed as much, if not more, by the extent to which a plant is pinched as by its natural habit. A tall dahlia for the back of the border is more suitable if the flowers tend to be gathered at the top of the plant, while a slightly shorter variety for the middle ground, perhaps fronted by a creeping or hummocky perennial, must not only be furnished with foliage as low down as possible but should also carry flowers on the sides of the plant as well as the top. Foliage is the last feature which must be considered before examining the flowers themselves.

### The Bishop

There is no doubt that dahlia foliage can be ugly: coarse in shape, heavy in tone and all too obtrusive. Fortunately there are, and indeed have always been, varieties with dark leaves. These have been described, over-enthusiastically, as red-leaved and purple-leaved; bronze-leaved is more accurate. It is in this group that many of the best border dahlias belong.

'Bishop of Llandaff', known in America as 'Japanese Bishop', is the most popular and widely grown. Alex Pankhurst has dug up its history and stored it for us in her book, *Who Does Your Garden Grow?*, a fascinating book whose title is its only flaw.

Raised by Fred Treseder at his nursery at Llandaff, now part of Cardiff, in 1924 it was chosen as the best of a selection of Fred's latest seedlings by the Right Reverend Joshua Pritchard Hughes, a non-gardener, a supporter of the temperance movement and a strict observer of the Sabbath – on which day he refused the assistance of any transport, preferring to rely on the two feet with which God had provided him for the purposes of locomotion. He selected the seedling from the dahlias presented to him by his friend, and his friend gave it his name, 'Bishop of Llandaff'. It received an AM at the RHS in 1928 and by 1936 the *Spectator* reported: 'The most popular flower of the moment in many parts of England in any test is the dahlia known as the "Bishop of Llandaff"'. It became so popular that seed companies sold packets of seed collected from it, under the original name, and here lies a lesson.

Fred Treseder tried more than once to raise other varieties from its seed. 'In one year we had 5,000 seedlings from this plant,' he reported, 'but none has excelled the parent.' So what did all those people who bought seed from Thompson & Morgan get? That this may be the origin of the

*Dahlia* 'Bishop of Llandaff' has enjoyed an enormous resurgence in popularity in recent years

various slightly different forms of the Bishop which we now see in gardens, differing mainly in the degree of flower doubling and tone of foliage colour, seems probable. Christopher Lloyd had a slightly different experience from Fred Treseder, but he leads us to the same conclusion: 'A batch raised from seed gives you some interesting variants from which to select the best (retaining them as tubers in future years) and discard the duds.'

In the garden the Bishop, occasionally known less respectfully as the Bish, is a vital red-border plant; whenever red borders are planned and planted it makes its indispensable presence felt, be it sneaking through the swords of browny purple forms of *Cordyline australis*, tussling with *Lobelia* 'Compliment Scarlet' or *Crocosmia* 'Lucifer' for dominance both in terms of colour and vigour, or perhaps with blue salvias or aconitums, so successfully used as a spark of contrast in red plantings.

There are other, equally valuable, dark-leaved

dahlias. 'Bednall Beauty' is a rather shorter version of the Bishop with the flowers in a darker red; 'Tally Ho' is also shorter and a little paler in flower, more vermilion than scarlet but with the same distinctive foliage.

Away from the Bishop's influence, 'David Howard' is one of the best in bronzed orange. 'Ellen Houston' has deep orange flowers although they sit rather tightly on the dark foliage, while 'Fascination' has very large, semi-double flowers which are held noticeably high above the foliage and are cerise, paling to the edges. Its leaves are especially rich in colour. 'Gaiety' seems rather weak, with double flowers in a slightly peachy pink. 'Preston Park' is a bold, single-flowered red but less penetrating in colour than the Bishop, while 'Yellowhammer' has single yellow flowers with a few slight orange streaks and superb dark leaves. 'Moonfire' is a single yellow with a red eye and is the most recent, introduced in 1993.

### The hardy dahlia

In recent years *Dahlia merckii* has become well known as the 'hardy dahlia', and if it can survive four winters on my clay I think hardy is a word well suited to describing its constitution. When it was introduced I cannot say, but Robinson mentions it enthusiastically in *The English Flower Garden* and it was Bowles's favourite; both knew it as *D. glabrata*.

It makes a great bushy plant, rather floppy in its second and third year and apt to collapse in a heap over its neighbours. The slender, almost translucent stems carry small, divided foliage and flowers in small groups at the tips. Individually the flowers are elegantly proportioned, not large, 1–1½in, but their six narrow rays create a star-like effect – or they would if they opened out flat; they usually nod and remain slightly closed.

In colour they vary. Robinson's deep purple I have not seen, though it must appear before long.

Various pale lilacs are more common, but like the stems their translucence adds a certain coolness. Pure white forms are around, and Nori Pope at Hadspen House has been working on this species in the last few years and produced the pure white 'Hadspen Star' together with a dwarf form.

### Cuttings, tubers and seed

Propagating dahlias used to be one of those subjects which gardening magazines enjoyed demonstrating in the 1970s and early 1980s. Step-by-step sequences of pictures appeared every year showing how to store the tubers over the winter, start them into growth in the spring, take cuttings and plant them out. Now magazines are more stylish; they concentrate on the look rather than the practicalities.

But the ironic thing, considering the pages of explanation and illustration which we used to flick past, is that there is very little to learn. You can buy tubers in the garden centre, you can order them from seed companies and from dahlia specialists; many dahlia specialists will send you rooted cuttings. Dahlias bought from non-specialists often come as plants established in pots.

It matters not the way you acquire your plants, but preparations for the following year require thought. The simplest and not necessarily the most expensive method is simply to re-order tubers or rooted cuttings each year. If your facilities are limited and your finances rather the opposite, this is the best course. But most of us will want to keep the dahlia tubers over the winter in preparation for the following season.

Lift them when the tops are black, cut down the tops, tie a label to the crown, hose off the soil, dry them on the greenhouse bench, then store them in orange boxes of dryish peat or vermiculite, with a few slug pellets thrown in. Place the boxes anywhere frost-free but not warm – the garage, the spare room, under the staging in the cold greenhouse.

In spring you have a choice. You can replant the original tuber in the garden in May; this will give you a big, bushy but late-flowering plant. Alternatively pot up the old tuber – it may need an 8in or 10in pot – and start it into growth, keeping it frost-free, before planting it out in early June with plenty of top growth. This will give you a big, bushy, earlier-flowering plant, but it may prove difficult to manage in the greenhouse if conditions are too warm; frost-free but as cool as possible is ideal, even in May. Shoots may need pinching to prevent them becoming lank and weak.

Short cuttings can also be taken from the youngest shoots as they first sprout from the tubers; these are rooted in a propagator, potted up, pinched and planted out in early June from $3\frac{1}{2}$in pots. These will give strong plants but not as big and bushy as those from tubers. The tubers can be planted out too and will still make good plants in spite of, or perhaps because of, all that pinching for cuttings.

There is one old trick which the enthusiastic dahlia grower can use to avoid the overwintering tubers taking up too much space. When planting out rooted cuttings, two of each variety are potted on into 5in pots and kept in a frame. They are kept there all summer, watered well, and the pots turned every week to prevent rooting through. If you forget to turn them, the daft dahlias will make tubers in the gravel under the pot. In the autumn the top growth is cut away, and the pots are stored on their sides on a rack under the staging of a frost-free greenhouse. Inside each pot will be a small tuber, quite big enough to provide cuttings the following spring.

The plants in the garden are just left there until you have a spare moment to dig them up and sling them on the compost heap. This method enables you to save space in the greenhouse for other plants or to store a far greater number of varieties in the same space.

Finally, a word about dahlias from seed. A steady stream of new varieties is appearing and in the late 1980s the first dwarf single colour strains appeared. There is not a strain available which is as good as the named varieties raised from cuttings, although in their first year at least many are a great deal shorter.

But just occasionally, in the dark-leaved 'Redskin' and 'Diablo' strains or in one of the collarette types like 'Dandy', a single plant of exceptional quality will appear. This, of course, can be propagated from cuttings.

Dahlia seed is easy to germinate and quick to grow. Sow the dwarf mix 'Figaro' in a cold greenhouse at the end of April and the plants will be in flower at the end of July. Sow the others in a propagator in March, prick out into individual pots and they will probably be in flower when you plant them in early June. *Dahlia merckii* even self-sows – I have an especially large-flowered, cool, pale lilac form as a result.

# Anemones

'At this time of year, this dull time, this heavy August time, when everything has lost its youth and is overgrown and mature, the Japanese anemones come into flower with a queer reminder of spring. They manage, in late summer, to suggest the lightness of spring flowers.'

This August melancholy described by Vita Sackville-West is one with which I have been afflicted myself. There seems a sad lull, weeks of empty silence between the tumbling of petals from the soft old roses and the arrival of the first seed catalogues – which coincides with the first autumn asters and the best of the Japanese anemones.

During this eerie summer, when we expect gardens to be at their most eloquent, I often feel deaf to their rhythms and their colours; they seem at such a distance. In these weeks, I hear only the raucous shouts of the summer bedding and am alert to the need for quiet calm and harmony and for simple elegance.

Into this restless conflict 'Honorine Jobert' opens her simple whiteness, a lark in the clear air. The purity of the colour, cradling the cloud of yellow stamens which only serves to emphasize the whiteness, has a kind of aloof innocence. The tall slender stems wave gently, making the name of windflower, originally coined for the spring anemones, as appropriate for these, so different. While the flowers open in late summer yet carry their reminders of spring, the fallen petals catch in the broad, dark vine leaves presenting hints of autumn. Or, as Bowles put it in just one sentence, 'There is a charm in the simple form of a single Anemone that goes straight to my heart.'

## Enough of all this...

Vita Sackville-West said that these autumn anemones brought a queer reminder of spring, but they are so unlike either the woodland anemones or the Mediterranean sorts that this highlights the power of her imagination as much as her observation.

These Japanese anemones have tough woody rootstocks and at the same time most develop a running habit. Their leaves are broad and bold; their autumn flowers are basically five-petalled, although doubles of varying degrees exist; they may reach 4ft in height, and although petals may fly in storms, the stems are rarely beaten down; they may lurch collectively to one side, as a collapsing greenhouse which by some unseen miracle remains standing, but they rarely disintegrate like the asters into a heap of broken detritus. The resemblance between these plants and the woodland and Mediterranean anemones is a botanical rather than a horticultural one.

There are five species which make up this Japanese collective of anemones: *A. hupehensis*, *A. x hybrida*, *A. japonica*, *A. tomentosa* and *A. vitifolia*. At once we have a contradiction, for Hupeh could hardly be described as being in Japan and we then discover that *A. tomentosa* comes from Tibet and *A. vitifolia* from Nepal. It appears, though, that this quintet has been rationalized, for *Anemone japonica* has been subsumed into *A. hupehensis* as var. *japonica* and most garden forms of *A. vitifolia* now belong in *A. tomentosa*.

The pink anemone of cottage gardens all over the country is known simply as *A. x hybrida*.

160

'Honorine Jobert', a white anemone of supreme
elegance

This indispensable plant arose in the middle of the nineteenth century at the old RHS garden at Chiswick in London as a hybrid between *A. vitifolia* and *A. hupehensis* var. *japonica* and was soon followed by its unsurpassed white sport, 'Honorine Jobert', which occurred on a French nursery in the late 1850s. These two are seen more than all the others put together and so,

with a fine pink and a superb white, why do we need to grow any of the others?

Many gardeners do not, but there is a darker pink I would recommend, 'Hadspen Abundance', yet another plant raised by the prolific Eric Smith. There is also an unusually large-flowered pink, slightly darker than *A.* x *hybrida*, called 'Königin Charlotte', and there is 'Géante des Blanches', which grows over 6ft in height with downy stems and huge, single white flowers; there's also the more modest white 'Louis Uhink', which can have up to ten petals.

Then there are the more consistently semi-double varieties. Some are rather messy beasts, the natural simplicity of the flowers lost in a confusion of petals in different sizes. 'Prinz Heinrich' in deep rose and the white 'Whirlwind' are less of a mess than the deep pink 'Monterosa', which is better from a distance when the strength of colour is more apparent than the chaotic form of the individual flowers. 'Bressingham Glow' is a good rich colour but with at least three times as many petals as a single and, like 'Monterosa', better from afar. 'Lady Gilmour' is probably the most double of all, with very large pale pink flowers

Allocating these forms to individual species has proved difficult; to the gardener the important factor is that forms of *A. hupehensis* like 'Hadspen Abundance' and 'Prinz Heinrich' flower earlier than forms of *A. x hybrida*, which is where the majority fall.

Japanese anemones are vital constituents of autumn borders alongside salvias and the classier asters, but careful choice of chrysanthemums is necessary as some of their colours can prove uncomfortable as neighbours to pink anemones. These anemones also make good background plants to temporary summer plantings, their slender stems rising above *Salvia coccinea* or lime green nicotianas just as they pass their best. In such a situation, where the ground is regularly replanted, containing their questing roots is also less of a problem.

### Woodland anemones

'Oh dear me, in what fit words can one describe the Quakerish loveliness?' Such is E. A. Bowles, getting up steam in his eulogy to *Anemone nemo-rosa* 'Robinsoniana', still one of the most captivating of the wood anemones but more than occasionally usurped by as beautiful impostors. 'And the shape of it, and the grace of it, and the dainty poise of it!' There is no stopping him once he builds up speed. 'A. robinsoniana is worthily named, for its delicate, refined loveliness.' Now let me get this right. Bowles saw in William Robinson a delicate, refined loveliness...?

This most famous of wood anemones, spotted by Robinson at the Oxford botanic garden but originating in Ireland, is more slender than 'Allenii' and slower-growing, which can be an advantage in situations where it nestles with choice but equally reticent neighbours. Miles Hadfield described 'Allenii' as an improvement on 'Robinsoniana', which rather misses the point, for apart from the distinction in vigour, 'Allenii', named after James Allen of Shepton Mallet who selected it in the late nineteenth century, is darker in colour, a deep lavender blue with purplish red streaks on the backs; 'Robinsoniana' is almost white on the backs, perhaps with a cream flush. Curiously, having lost his heart to 'Robinsoniana', Bowles then falls for 'Allenii' and describes it as 'the largest and loveliest of the blue Wood Anemones'; we can reasonably presume Bowles to be warm-hearted rather than fickle in his affections.

These blues can be confusing, especially as it is hardly unknown to order one and be sent another. But 'Blue Bonnet', which migrated from Wales to Tom Smith's Daisy Hill Nursery at Newry and then crossed the water back to Britain, must have a mention. This is not only one of the most tolerant, both of poor soil and sun, but is also unusually late-flowering. The flowers are more solid and in a rich clear blue, and jostle around bold bergenias rather than yellow hellebores.

In more recent times, new anemones have

been selected. Paul Christian found 'Pentre Pink' in his own Welsh wood and collected two small pieces of rhizome from which to build up stocks, only to find that the plants left in the wild faded away. This is perhaps the best pink, and certainly better than the rather dirty 'Currie's Pink', while 'Rosea' has unusually coarse and heavy foliage.

The weird 'doubles' are craving attention but on the way 'Leeds Variety' must make an appearance; if Bowles devoted one and half pages to it and Farrer considered it to be the best of all we cannot but expect it to be a good plant. The flowers are unusually large, pure white and with a pink flush to the backs.

Now, 'Alba Plena', 'Bracteata', 'Bracteata Plena', 'Bracteata Pleniflora', 'Flore Pleno', 'Green Dream', 'Green Fingers', 'Monstrosa', 'Virescens', 'Viridiflora' – the names give you the general idea, but they are much confused, so an attempt at differentiation is in order.

'Alba Plena', which has been grown since Gerard's time, is much smaller than wild forms, with a ring of six pure white petals and a tight white button in the centre made up of stamens which have become petaloid; the form is much the same as that of *Ranunculus ficaria* 'Collarette'. 'Vestal' is similar only much larger. 'Bracteata' is double in a different way, making a ragged mophead of narrow white petals, some streaked with green, some entirely so. Bowles was unimpressed: 'There is a mild excitement to be obtained from growing such an unreliable plant and watching its vagaries...' Others find the fact that in some years it can be single and in others fully double of rather more interest. Robinson seemed to muddle it with 'Alba Plena'. Plants labelled 'Bracteata Plena' seem to be no different, while 'Bracteata Pleniflora' also seems the same.

'Flore Pleno' seems much the same as 'Alba Plena' but with a frillier centre; but then there are the curiosities. 'Green Fingers' has little white flowers which fade to pink but with the strange addition of a tuft of miniature green leaves in the centre of each flower. Some gardeners are very rude indeed about 'Monstrosa' and it must be said there are more elegant forms. But with its petals which have become partly leafy and dissected, it presents a strange picture. 'Virescens' and 'Viridiflora' are probably the same thing, with all the floral parts transformed into miniature leaves, making a mound of green lace. This is quite captivating and also seems a strong grower. It bewilders the uninitiated, which is always a pleasure. The new 'Green Dream', in green and pink, I have not seen.

I confess to being intrigued by the hybrids between *A. nemorosa* and the yellow *A. ranunculoides*. There are two, sometimes known as *A. x lipsiensis* and *A. x seemanii*, but current thought favours viewing both as forms of *A. x lipsiensis*, 'Pallida' and 'Seemanii'. The former has brighter flowers and greener foliage and is the more robust, 'Seemanii' is paler and has bronze leaves which are especially striking as they emerge early in the season. It seems to me that there must be more scope for hybrids here, if anyone can be bothered to make them. Seed can be germinated without difficulty if sown fresh, and once individuals have been selected and allowed to multiply they can be lifted in autumn and the rhizomes broken up and replanted.

And *A. ranunculoides* itself, with bright buttercup flowers over delicate foliage, establishes quickly, increases well and is often so voluminous as to be better set back among shrubs than as a frontal plant where more retiring treasures may be in danger of suffocation. There is a very pretty double, 'Pleniflora', and a form which I have not yet seen, var. *laciniata*, with unusually finely cut foliage.

All these anemones are easy to divide but are

often unsuited to long periods of pot cultivation; the rhizomes need room to run.

### The spring carpet

Although some wood anemones like 'Robinsoniana' tend to look warily upon fresh pastures and remain compact, many, and especially perhaps 'Blue Bonnet', spread well and thickly to create the perfect background to specimen plants. This particular variety may be too late in flower to creep under the skirts of 'Queen of the Night' hellebores while it's still at its best; 'Allenii' would be a better choice. But I once saw it wandering among the suckering stems of *Prunus tenella* 'Fire Hill', lined with pink flowers, to create a startling effect. It looks good too with the fresh green shuttlecocks of *Matteuccia struthiopteris* surging through it or the rusty new fronds of *Dryopteris erythrosora*.

A variety like 'Viridiflora' demands, and repays, close inspection and is most enjoyable when planted with others which are best appreciated from a kneeling position. Elizabeth Strangman's extraordinary double *Helleborus torquatus* 'Dido' perhaps, with green and purple tints in its delicate flowers, or *Arisarum proboscoideum*, whose leaves you can part, while scrabbling about on all fours, to find the mice and their tails hidden underneath.

On a grander scale, 'Blue Bonnet' or 'Leeds Variety' will make impressive rolling blankets under deciduous shrubs in any reasonable soil, especially if mulched every autumn. They look good around *Magnolia stellata* and its darkest pink form, 'Rubra', which if grown well sweep right down to the soil. Camellias, *Amelanchier lamarckii* with its bronzed young foliage and, I suppose, rhododendrons also look good sailing on a blue or white sea of anemones.

There are two other species which fall, with *A. nemorosa*, into the category of more or less rhizomatous woodlanders, *A. blanda* and *A. apennina*. Most forms of *A. blanda* are more forceful, in colour and substance at least, than those of *A. nemorosa*. 'Radar' is an extraordinary almost magenta pink, which needs very careful placing – next door's garden perhaps. 'White Splendour', by contrast with wood anemones, may be large and pure in colour except for its blushed buds, but its vigour and undeniable dominance can be welcome under shrubs.

*Anemone apennina* is more delicate in habit and has fewer variants. Soft, pale blue is perhaps the commonest shade, although there are also white and less common pink forms. This species will naturalize happily among border salvias or hardy fuchsias, creating a soft carpet early in the year, fading away as the later flowers come to dominate.

### Mediterranean anemones

Mediterranean anemones require contrasting conditions. Mine grow in a south-facing raised bed filled with rich yet well-drained soil and increase steadily. Four species fit here, *A. coronaria*, *A. hortensis* and *A. pavonia* plus *A. x fulgens*, the hybrid between the latter two. The cut flower anemones of the de Caen and St Brigid groups are usually placed under *A. coronaria*, with the St Bavo group coming under *A. x fulgens*. Unfortunately these, like most groups of anemones, seem plagued by confusion over their names and to which species they should be assigned.

But in all these species there are superb, rather less developed plants, often derived from wild collected stock, such as the pink and smoky forms of *A. pavonia* sometimes offered, together with a spectacular double crimson resurrected and distributed in modest quantities by Graham Gough. It would also be intriguing to recreate some of the old fully double, flaked and picotee forms, and the Hardy Plant Society is encouraging work along these lines. And, occasionally, plants with a simple, unaffected look turn up in the de Caens.

In the south of England, given the conditions I describe, these Mediterranean types increase well, hold their flowers boldly towards the sun, and the singles at least set plenty of seed from which appealing shades can be sometimes selected. With *Mahonia fremontii* as background and small cistus as wind protection, they make a little corner in a Mediterranean style. In cooler and damper regions, protection will be necessary.

# Asters

'Gardeners,' says Mrs Fish, 'are divided into two classes, those who grow Michaelmas daisies and those who don't.' Well that puts me in a pickle, for I fall into both camps – and for two different reasons! First of all, it is only in recent years that I've come to appreciate any of them. I somehow thought of them like Charles Dickens, a worthy pleasure which had inexplicably passed me by. Then two things happened to open my mind to their attractions. First of all, Elizabeth Strangman and Graham Gough at Washfield Nursery started enthusing about them; and when a group of plants captures their attention, astute gardeners take notice. They talked about the true 'Climax', about rediscovering an old pink *A. amellus* called 'Jacqueline Genebrier', and about 'Monte Cassino', a cut flower from Holland.

Soon after, the RHS embarked upon a large trial of *Aster* species and hybrids at Wisley which included almost any aster apart from the many forms of *A. novi-belgii* and *A. novae-angliae*. With almost any group of plants the discipline of having to examine them carefully and intimately reveals their special qualities. Here were all the confusing forms of *A. x frikartii*, the dainty clouds of *A. ericoides* and *A. cordifolius*, plus possible and probable hybrids. I looked and learned and enthusiasm continued to surge; and once I'd spotted the wonderful 'Ochtendgloren' there was no holding back. A vast undiscovered territory was opened up and the gold rush began.

For what was so startling was that so many of these plants were so good. As William Robinson put it: 'But the most precious, perhaps, of all the flowers of autumn for all parts of the country, grouped in an artistic way, are the hardy Asters of the American woods ... no such good colours of the same shades have ever been seen in the flower garden.' He goes on to commend 'the still more precious Asters of Europe, which, by their extraordinary beauty, make up for their rarity'.

Christopher Lloyd is understandably more selective and has abandoned the many varieties of the mildew prone *A. novi-belgii*: 'I am quite happily giving up growing these plants altogether,' he says. 'Mildew gets worse on them from year to year and verticillium wilt is a frequent and lethal disease among their ranks. They flower for a very short time in autumn and look terribly boring all through summer. So, one way and another, I have few regrets.'

Others he does grow, including the Lloydianly lurid *A. novae-angliae* 'Andenken an Alma Pötschke' and many other mildew resistant types. Perhaps a visit to Paul Picton's aster collection at Old Court Nursery in Worcestershire might encourage him to try again some of those lovely old varieties. But why should he? He looks forward rather than back and indeed in his book on perennials from 1967 he recommends that plant breeders develop other kinds of asters. Some of their successes could be seen in that RHS trial.

### New Belgium and New England

For many gardeners, myself included until a few years ago, the very phrase Michaelmas daisy was a warning. This may have come about because, like rosebay willowherb, seedlings of *Aster novi-belgii* colonized city and suburban railway embankments and sidings; everyone knew rosebay was a terrible weed, so Michaelmas daisies must be too. They can be weedy, both in their individual growth and in their tendency to spread, but this combination of tattiness and enthusiasm can hardly be called intolerable.

This prejudice is reinforced by a susceptibility to powdery mildew, which is perhaps their most irritating fallibility. Fortunately, powdery mildew is no incurable plague; if you spray your roses, mix up a double batch and spray the Michaelmas daisies too – end of problem. But, of course, prevention is better than cure and this can be undertaken in two ways. Powdery mildew thrives in hot, dry conditions; so eschew the *A. novi-belgii* group if you garden in an area where hot, dry summers are common. The second form of prevention is simply to mask the damage; mildew usually kills the foliage on the lower half of the plant without disfiguring the upper part or disrupting flowering. So the answer is to plant something shorter and bushy in front to hide the bare stems.

I, or rather my plants, have never suffered from the tarsonemid mite which plagued the asters at Dixter and to a limited extent in that RHS trial in the early 1990s. None of the chemicals available to home gardeners will deal with it, so vigilant inspection of new plants is the best prevention.

Having insisted that you grow these asters, which in spite of their drawbacks require no more special attention than many other plants, how is it possible to choose from so very many? Over 200 varieties from the Old Court collection can be propagated for sale. It is less difficult

when the search is approached from the opposite direction to the usual – that is, not by glumly saying: 'There are so many which are good, how can we choose?' Instead, let us work on the assumption that all but the very best forms of *Aster novi-belgii* will be overlooked. We will consider only the few that are so special we simply cannot ignore them.

Visiting Old Court to see the collection, I saw exactly why borders devoted entirely to Michaelmas daisies were once so popular. The richness of their colours harmonizes so well with colouring trees round about, but more importantly they harmonize with each other.

Having spent a few hours looking at the collection, certain characteristics come to the fore as especially significant, aside from the actual colour. Apart from susceptibility to mildew, the colour of the stems and the extent to which the colouring extends into the flower head are worth looking at. Also, the speed at which the eye of the flower turns from a clear to a dirty colour; sometimes this takes place even before the flower is fully open and can spoil the look entirely. Another factor is the degree to which the ray florets curl up; the flowers of some varieties hardly open properly at all. And are the plants self-supporting? Do the faded early flowers detract from the later ones?

The first of the five I selected at Old Court is 'Climax', which Graham Thomas says 'has never been surpassed for vigour and elegance'. This is a tall, genuine October-flowering variety with fresh-looking, slightly glossy foliage right down to the soil. The buds are a soft purple, opening to elegant open heads of pale lavender blue; the eye yellows increasingly as the flowers age. The combination of the excellent foliage, elegant flower head and clear colour makes this indispensable. 'Dauer Blau' is half the height, with dark stems and a dramatic head of lavender blue flowers whose rays darken at the base and pale at the tips. The yellow eye fades to a more gingery shade.

'Goliath' is aptly named indeed, resembling nothing less than a 6ft version of *Aster* x *frikartii* 'Mönch'. The misty purple buds open to soft lilac blue with a green eye, soon yellowing to buttercup. It has a similar elegance to 'Climax', although as the flowers age they do become a little messy. 'Harrison's Blue' at 3–4ft is the best deep colour, but more of a silvery purple than a blue, with the flowers held on short stems and cradled in rather mossy bracts. The eye slides from orangey yellow to a more toffeeish shade.

'Marie Ballard' is, I think, the only one of Ernest Ballard's many varieties to make this selection. At 3ft this full-petalled double blue with a touch of lilac makes a very neat, clean-edged circle of rays; the eye only emerges in the last stages of each flower's life. The foliage is good, the stems too being purple except in the very tips of the flowering shoots.

I might add to these 'Coombe Rosemary', a good pale purple double which retains its fullness for some time and keeps its dark, slightly bluish foliage well. Also the 18in, purple-stemmed 'Kar-minkuppel', with pale rose flowers and green eyes which fade happily to a watery raspberry.

A good white is most conspicuous by its absence; whites usually fail through their old flowers detracting so determinedly from the new by curling up and turning brown. Miss Jekyll spotted the same gap in these apparently well-filled ranks and also found the soldier to fill it. 'I always group with the Michaelmas daises the handsome tall white daisy, *Pyrethrum uliginosum*. Though it is a plant of a different family, it is of daisy form and flowers with the Asters; and as there is as yet no very large white flowered Aster, it answers to the need in an admirable manner. Indeed I am not at all sure,' she continues, 'that it will not always keep its place as the most suitable companion to the Michaelmas Daisies, for it must of necessity be a long time before a white Aster can be evolved that can come into competition with its hardy nature and bounty of large white bloom.' Now known as *Leucanthemella serotina*, it still keeps its place, for nearly 100 years later no competitor has indeed evolved.

At Old Court these New York asters look good in borders of their own or with other types of aster. They make a wonderful autumnal herbaceous planting in full sun with *Sedum* 'Brilliant' or 'Meteor', *Schizostylis* 'Jennifer' or 'Major', and hardy chrysanths like 'Bronze Elegance'.

William Robinson grew them in a very particular way, and one which we do not see now. Indeed he is rather snooty about having developed a superior way of growing them: 'Many of these, of an inferior order of beauty, used to be planted in our mixed borders, which they very much helped into discredit,' he insists. But he has discovered the answer: 'The best of these massed and grouped among shrubs or young plantations of trees, covering the ground, give an effect new and delightful, the colour refined and charming, and the mass of bloom impressive in autumn.' He grew them in their thousands, covering clearings and glades in what must have been a stunning spectacle.

This is impossible now, it was impossible for most people then, and I still like the idea of an aster border such as can be seen at Old Court – there are a few varieties which maintain sufficient interest for long enough to make this a not entirely foolish notion, *Aster lateriflorus* 'Hori-zontalis', for example. But even this outstanding plant is perhaps best combined with other autumn plants – hardy chrysanths, schizostylis, ceratostigma and so on.

Propagating this group is perhaps too simple. Dig them up; with the two border forks used in the traditional manner the roots fly apart; back go the pieces. No. Wait until spring for a start, just to be on the safe side. Lift the clump, snip off the fattest juiciest pieces from the edge, chuck the rest away or take shoots off as cuttings (they will root easily enough), then replant those plump, rooted shoots and they will romp away. Sometimes doing this every year is not too often; every other spring is ideal. The worst thing to do is detach pieces from the edge, pot them up and give them away, leaving the tired old clump in place. It may fade away entirely.

*Aster novae-angliae*, asters from New England, we all know where that is, are indeed different from those from the lost continent of New Belgium (which turns out to be modern-day New York). To continue with propagation for a moment, the roots of this group are more woody, more compact and come to pieces a little less easily. But the same rules apply; young shoots off the edge, or cuttings, are all you need. With both groups dead-heading is necessary or seedlings, probably in quite another shade, will germinate in the clump.

The New England asters rarely seem to be troubled by mildew – although this does not prevent simple drought frazzling the lower foliage. Their foliage is broader, softer and paler – which is sometimes an enhancement, sometimes not, but I find them strangely ungainly. Unfortunately, in bad weather the flowers remain half closed through the ray florets standing upright instead of lying out flat; this spoils them.

There are far fewer varieties in this group and all are much the same height and type. 'Andenken an Alma Pötschke' is an excessively lurid shade of pink but would be suitably startling with yellow chrysanths, although the tips curl up and go frizzy. 'Barr's Blue' is not blue but pale violet; it dies a horrible tortured death as the tips crisp and curl up quickly and the eye turns dark and rusty all too soon, so that's that. The sugary pink of 'Harrington's Pink' is one of the few shades difficult to accommodate in the autumn garden, and 'Rosa Sieger' compounds the problem with its larger flowers, although they open almost red. 'September Ruby' is not ruby but a sort of magenta cerisey pink, while the brown eye of 'Treasurer', sometimes known as 'Hella Lacy' in the States, detracts from its silvery-backed lilac flowers.

When I first started to think about New England asters I hardly intended to be so harsh. But something about their stiff habit and their determination to lose the bottom third of their foliage even in good seasons somehow infiltrated into my subconscious and nudged all the niggling irritating points about the individual varieties to the surface.

So let us be honest. I have lived without the New England asters for years and could happily continue to do so. I could live without most, but not all, of the New York asters; but those in the next two groups are essential.

### Aster amellus *and* Aster x frikartii

'A large plant of rich blue Aster amellus, covered with peacock butterflies, is a common sight enough,' says Canon Ellacombe, 'but it is a sight to be thankful for.' And I thought a man of the cloth might express a little more enthusiasm for one of the more spectacular sights of his God's creation. 'Damned with faint praise' is, I believe, the phrase. Mr Bowles is more fulsome: 'An ancient clump is often a very fine sight, all the flowers being packed so closely they form a level mass of bloom, and make a platform for the Admiral butterflies to alight on to suck their nectar.' He had a splendid idea for growing them too: 'If I had room for it [and how many times have we said that] I should like a long, narrow border facing south in front of an old stone wall filled with every variety procurable of *Aster amellus*. One might grow spring bulbs underneath them, but nothing that flowers or wants room for leaves later than June.'

This must go on the list of ideas for the new garden. Unfortunately it features at the foot of the third page; at the top of the first page is written: 'No overambitious schemes, do a few things well rather than lots of things badly.' Well, I can try...

In the wild *Aster amellus* is a rather variable species, hardly surprising when it is found wild in both Lithuania and Macedonia and much of the territory in between. As Bowles points out, it has a habit of carrying its flowers in a plateau of shimmering purple, and although sometimes starting to flower a little early, the simplicity of these sparkling discs woven together into a rich carpet takes the breath away.

But even among these essential plants some are a great deal more essential than others. The blue-tinted lilac 'Moerheim Gem' flops, even though only 15–18in high; 'Pink Zenith' is not pink, while 'Jacqueline Genebrier' is certainly pink, and self-supporting too, but the rays go blotchy in wet weather; 'Lake Geneva', in smoky blue, just seems to lack any definable character.

At the opposite extreme, the lilac blue 'Vanity' is one of the best of all autumn flowers for butterflies but the stems tend to sway outwards and then be battered by rain. 'Brilliant' may be a pink with a carmine tint but it rarely seems to go blotchy. 'King George' combines 3in flowers with 2ft self-supporting stems and is another

especially popular with butterflies. 'Violet Queen' resembles nothing less than a huge, 18in violet brachyscome.

Around 1920 Frikart's nursery in Switzerland raised three seedlings by crossing *Aster amellus* with the uncommon *A. thompsonii*. Perhaps they got the idea from the Reverend Wolley Dod, who exhibited, and then lost, such a hybrid in 1892. They named their seedlings after three famous mountains: 'Eiger', 'Jungfrau' and 'Mönch', and then later introduced another called 'Wunder von Stäfa'. 'Mönch' and 'Wunder von Stäfa' have made all the running – to the extent of becoming hopelessly confused in the course of the race.

Both these plants are unusually long-flowering, and the one which generally seems to be regarded as 'Mönch', but is not necessarily distributed as such, is distinguished by its distinctive flat flowers; the ray florets stand out almost perfectly flat from the disc (unless badly battered), with few above or below the general line. It is a clear blue and reaches about 3ft in height. 'Wunder von Stäfa' is a foot taller, slightly less self-supporting, in colour less blue and noticeably more purple, but perhaps most noticeably the rays are less evenly flat; some are set at a slight angle above the rest, some below, giving the effect of an altogether more ragged flower. This is the plant usually seen, whatever its label says.

The one which shocked me in the Wisley trial is 'Flora's Delight', raised by Alan Bloom in 1954 by crossing the pink *A. amellus* 'Sonia' (itself raised by a man with the splendid name of Mr T. Bones) and the dwarf form of *A. thompsonii*. The plant now seen under this name is worthy of nowhere but the bin. I can only assume that some interloper crept into Graham Thomas's garden under the guise of 'Flora's Delight', otherwise he would hardly have been prompted to call it 'one of Alan Bloom's best efforts'. The plant entered into the trial by Blooms was weak, scrappy, has puny flowers, starts flowering before just about every other aster, in July, and has seed-heads which soon detract from the display. It hardly deserves to be considered in the same chapter as Frikart's hybrids.

Unlike some asters these plants do not require annual propagation; splitting into small pieces in March or April every three or four years will suit them very well. But wet soil will not, and in heavy conditions the addition of drainage material or building up the bed using a raised edging will be needed to prevent them languishing in winter wet.

### Small-flowered asters

I have to say that with the exception of the *amellus/frikartii* group and a select range of the *novi-belgii* varieties my inclination lies most towards these smaller-flowered varieties. This is not to say that they are less colourful or less showy than all the others, more that they manage their display in an airier, less forceful and less provocative manner than do the *novae-angliae* types in particular.

Many of these small-flowered varieties are derived from *A. ericoides* ('does not belong in anyone's garden,' says Claude Barr in *Jewels of the Plains*) and *A. cordifolius*, although other species are sometimes involved. The Wisley trial of these varieties, apart from revealing a certain confusion among the names, showed us some great wonders.

'Golden Spray' was the best yellow, for although the rays were creamy, with its yellow eyes it certainly produced the strongest shade.

Three of the best asters (*left to right*): 'Herfstweelde', *Aster lateriflorus* 'Horizontalis' and *A. novi-belgii* 'Climax'

'Hon. Vicary Gibbs' in pale blue was slightly spoiled by the old flowers turning very brown but was certainly generous. The best of all the blues, 'Photograph', was wonderful at Old Court too. With 'Ochtendgloren' this was the best in the whole trial, which initially ran to over 100 entries. The rays of 'Photograph' are a pure, pale grey-blue and fade not at all until they suddenly go brown after hanging on for as long as possible, while the eyes are lemon and retain their colour very well before developing slightly pinkish tones. It has an elegant, slightly arching habit, rather like a firework bursting from an explosive centre.

The other real stunner was Piet Oudolf's 'Ochtendgloren', a cross between 'Monte Cassino' and another form of *A. pringlei*, which has one feature which seemed to set it apart. As the season progressed the clump continued to instigate new shoots at its edge, so that by flowering time the earliest and tallest shoots were 5ft high and the most recent just 12in; then they all flowered together. The result was a curtain of pink stars from top to bottom. And this reminds me of a difficulty, for while the flowers are identical to those of 'Pink Star', 'Ochtendgloren' is slightly spreading in habit while 'Pink Star' is determinedly upright.

It is not difficult at this point to enthuse also about the pale blue 'Herfstweelde', another from Piet Oudolf, 'Monte Cassino', *A. lateriflorus* 'Horizontalis', *A. paniculatus* 'Edwin Beckett', 'Ringdove', not to mention 'Little Dorrit' and 'Little Carlow', 'Cinderella' and 'Erlkönig', plus 'Kylie', a pink hybrid between *A. novae-angliae* 'Andenken an Alma Pötschke' and *A. ericoides* 'White Heather'. But I must stop.

### Out with the neck, axe at the ready

It has been suggested to me, after rambling on about these plants, their attractions and their various failings, that I should make a choice, that people read a book like this for specific recommendations. They want it laid on the line: 'Plant these, they're the best,' I must say. I wondered about giving a top ten for every genus I discuss, or perhaps picking just one star; in the end I decided to give a top ten, but only for asters, weighed down as they are by an unusually large burden of varieties.

So here, in alphabetical order, is my choice of ten asters which every garden should have.

Searching carefully at Bressingham on a number of occasions, I consistently scored *Aster amellus* 'Vanity' as the aster most attractive to butterflies, moths and bees, and it should rate highly for that alone. But its large lavender blue flowers, although not perfect in form, come in such prodigious quantities that it gets into the top ten in spite of being a fraction floppy. Choose 'Violet Queen' if you refuse to stake.

*Aster cordifolius* 'Photograph' is a superb plant with dense clouds of tiny grey-blue flowers, while the fresh-looking foliage of *Aster ericoides* 'Golden Spray' helps make it the best of the yellows, the colour coming more from the large golden eye of each flower rather than from its creamy rays.

The excellence of *Aster frikartii* 'Mönch' hardly needs repeating, its long season from June to October adding to the elegance of its straight-rayed flowers and its cool colouring.

*Aster lateriflorus* 'Horizontalis' is rather different from most of these asters in that the purple-tinted foliage and the stiff, angular habit of the plant, which make their presence felt for so long, are as much of an attraction as the small, white-eyed, pink-rayed flowers.

Just three representatives of the vast crew of New York asters, two of which are among the less common. *Aster novi-belgii* 'Calliope' has been reintroduced in recent years and stands out for its tall, shining, deep purple stems, branched from about half-way up, and its lilac blue flowers with eyes opening yellow-gold then eventually becoming ginger-tinted.

*Aster novi-belgii* 'Climax' is a genuine October aster and one of the most famous of all, though hardly the most widely available. Its open, pyramidal flowering head marks it out along with its clean foliage, and the soft purple buds open to clean pale lavender flowers.

*Aster pringlei* 'Monte Cassino' was grown for some years as a cut flower, in Holland especially, and is still forced for the markets; I've seen it sold in December and May. Small white daisies in tall airy wands make it a superb border plant as well. I seem to recall enthusing about *Aster* 'Ochtendgloren' already.

# Latecomers: A Final Choice

I have given the season's major players, the Japanese anemones, asters, chrysanthemums and dahlias, chapters of their own and could probably have done the same with twice as many plants. But here I choose just one further group so necessary to the autumn garden, then conclude with a dash through this undiscovered season's discovered and undiscovered plants.

### Autumn composites

So many of autumn's plants are members of the daisy family that it seems indulgent to discuss even more than the three covered so far at the expense of others; in fact I could have given the others the whole of this chapter but choose to mention only boltonias, eupatoriums, heleniums and helianthus.

Few gardeners in Britain have even heard of boltonias and perhaps this is because although we grow four out of the five available, the best has only just arrived in a British garden – in mine, to be precise. I rather suspect that at present the twenty nurseries which sell them are very much ahead of gardeners in their awareness.

The clouds of white daisies on 'Snowbank' are delightful and provide a different style from asters and chrysanths for early autumn. 'Pink Beauty' has a much longer and later season and is more valuable for that single quality alone, but the flowers are also larger, the foliage a bluer green and of course the colour is pink. Both have the additional delight of the fading foliage developing valuable gold tints as the season closes. 'Pink Beauty' was found by Edith Eddleman, who is in

charge of the perennial border at North Carolina State University. It should be on sale in the UK soon.

In small gardens eupatoriums, known in the USA as bonesets, may be difficult to manage, for many are uncomfortably large there is no doubt, but the nine pages Allen Lacy devotes to them in his inspiring book *The Garden in Autumn* are testament to their value. For the sake of gardeners with plots of modest size it is convenient to separate off the white-flowered from the generally much taller and more invasive purple-flowered species.

Bowles thought the white-flowered *E. ligustrinum* the equal of *Sedum spectabile* as red admiral bait, but from my observation it is good, but not that good. Over neat, bright green, nettle-like foliage the flat white heads resemble those broad white ageratums grown for cut flowers; although perhaps a little less sharply white. The white snakeroot *E. rugosum*, also known as *E. ageratoides*, is so similar the distinction evades me. *E. coelestinum*, sometimes called the hardy ageratum, looks to be just that, in a similar fluffy blue but with the flowers on 3ft plants.

All make attractively loose, rounded clumps in contrast to the more starkly stiff or upright habit of most asters and many other autumn specialities. *E. ligustrinum* should be in every autumn border.

The taller, generally more coarse, purple-flowered species are for the back of the border and larger than average ones at that. In the States *E. purpureum* is known as Joe Pye weed; I cannot tell you for certain why. Allen Lacy gives two possible theories: 'One holds that in colonial times an Indian medicine man named Joe Pye cured typhoid fever and other ailments with a ghastly tasting tea brewed from the plant's leaves. The other theory is that in one of the American Indian languages jopi was the word for typhoid. Whatever the case, Joe Pye weed has no place in modern medicine.'

This 8ft monster certainly has a place in larger borders; its huge, bold, branched heads in a sort of rosy purple fit so softly into the autumn scene. For so substantial a plant to have such a quiet presence in the autumn garden is something of a blessing. The far more feathery dog fennel, *E. capillifolium*, looks to belong in a different genus altogether, resembling more a sort of purple, heavy-duty miscanthus.

In Britain *E. cannabinum* is a wild weed which is occasionally grown in gardens, although the double form is a better bet, for not only does it not seed itself, hooray! but it flowers for longer. Don't waste time smoking the dried leaves, they only look like cannabis.

Next, a pair of yellow daisies, those much maligned plants the best of which are such special autumn features. It is true that many of the earlier August varieties must be excluded here, but more for their timing than their qualities. They are coarse, they are, so refrain from planting them by the front gate or the path to the shed where this will more easily offend. Use them at the back of the border.

Heleniums, the sneezeweeds, hardly belong here, for they are more correctly late summer rather than truly autumn plants, they overlap from one to another; the American writer Helen Van Pelt Wilson calls them transition daisies, cusp daisies is good too. They always look as if they've become slightly disarranged, hurrying to the party, and are without a friend to take them to one side and suggest they find

a moment to compose themselves.

So although even their brightest yellows have an autumnal air we must pass on to *Heliopsis*, but then most of these fall into a similar transitory phase, although *H. helianthoides* (that means 'the heliopsis that looks like a helianthus', I'm sorry to say) is certainly later to flower than *H. scabra* and so deserves a moment's attention. But now I find that everyone has bundled all the heliopsis into this one species, and *H. scabra* is reduced to a subspecies. Rarely grown in Britain, the late one in which we are interested has semi-double flowers with yellow centres.

Finally, *Helianthus*, not the annual sunflowers although these can look splendid in the autumn when frosts hold back from reducing them to stark sticks. No, there are a few species which are genuine perennials and genuine autumn plants. In the States they seem to be more widely grown than in Britain.

'Of all the fall-blooming perennials, I prize most highly this sunflower,' says Allen Lacy. To which sunflower is he referring? I hear you ask. You won't guess, especially if you read this in Britain; *H. angustifolius* is the answer, a plant not mentioned by Graham Thomas or, according to *The Plant Finder*, stocked by any nursery in Britain. It gets a mention from good old Bill Robinson, but he gave these sunflowers over two whole pages in my 1906 edition of *The English Flower Garden*. Actually, as is so often the case, he did not actually write the entry himself – one D. Dewar was cajoled into making a contribution – but they are still allowed more space than border phlox. 'The Sunflowers, like the Michaelmas Daisies, could ill be spared from the autumn garden,' says Mr Dewar, 'where, when most other hardy perennials are beginning to show the sere and yellow leaf, they are generally at their best and in their greatest numbers.' Exactly, but how times change, for Graham

Thomas is so much less keen: 'I cannot write about these with any enthusiasm.'

I have not grown *H. angustifolius* and neither have I ever seen it, but pictures and reports convince me that it is at least worth trying. Another I have neither seen nor grown is *H. giganteus*. 'Tall and elegant' according to Mr Dewar, 10–12ft high or more, and Canon Ellacombe is equally enthusiastic, pointing out that it will reach 10ft in good soil 'and produce for many weeks an abundance of flowers of the purest yellow'. Why do we no longer seem to grow it? We obviously once grew it widely. 'It seems to be very popular for the gardens of railway stations, in several of which I have seen it in great abundance,' says the Canon. That must be the answer: first the nationalization of the railways in 1948, then Dr Beeching and his cuts, now with the privatization of British Rail the last plants will doubtless be lost to British stations for ever.

Finally, having scrupulously enthused about two plants I have never even seen, I come to one which I have and which I can recommend with confidence. I can tell you it is neither demure nor restrained and certainly not diminutive, for *Helianthus salicifolius* too can reach 12ft in height. Fortunately it is of interest long before it reaches its maturity, for as it grows its foliage (*salicifolius* = with willow-like leaves) falls from its shoots in a delightful outpouring of narrow strands, a little like some of those surprisingly elegant fountains seen in shopping centres. So it looks good as it grows and then in October there's a bold spray of bright yellow daisies at the top. I cannot deny it, this species needs the right spot. The angle in the corner at the back of the border may well be the place, for although the fountain will not be revealed until the summer it will be all the more impressive for its unexpected arrival.

### And finally...

Simply by turning to autumn with an optimistic gaze rather than a glum apprehension over the chilly wait until spring, an ever-increasing range of fine perennials will be noticed and enjoyed. I'm convinced that a true appreciation of autumn depends on open-mindedness and a determination to pass on, to look down from the autumn spectacle of the maples and oaks.

Not that perennials themselves are unable to provide autumn foliage colour, hostas and hardy geraniums often give a final flamboyant fling of colour before collapsing. And the uncut stems of composites are dusted with winter's first rime in the gardens of those less than scrupulously dedicated to following traditional recommendations to cut and cart away.

But take a day off from the insistent flamboyance of the maples, the liquidambar, the fothergilla and the euonymus and take a stroll around a garden where some attention, nay imagination, has been given to autumn plantings. You will doubtless find those vital genera I have already discussed but, surely, you will also come across aconitums.

Most are truly flowers of summer and may linger, but one is an autumn speciality, the latest into flower, *Aconitum carmichaelii*, and more especially its forms 'Barker's Variety' in a clear bright blue and 'Kelmscott' in a more violet-tinted shade. Both can reach 6ft, both have excellent foliage so they make their presence felt well before flowering, and their distinctive style and colouring fit well with other autumnals.

Some gardeners find it difficult enough to believe that there are any hardy begonias at all, but then to insist that they are stalwarts of the autumn garden seems to stretch their credulity a little too far. But *B. grandis* var. *evansiana* is a fine plant, with pink-and white-flowered forms. Both reach about 18in in height and are well furnished with bold, pale, brightly red-backed leaves down to ground level; the last lingering flowers of *Tradescantia* 'Isis' peeping through add a special blue sparkle. True, they will not survive well in cold, claggy soil or where the ground is regularly frozen for weeks on end, but in rich but well-drained soil, and watered well in summer, this species will overwinter contentedly in many parts of Britain.

Here, I cheat. For slipping in the ceratostigmas is perhaps a little sneaky, but I like them so much and they are so valuable for their habit as well as for their flower and foliage colour that I immediately forgive myself the breaking of the rules. For they do tend to keep at least part of their woody top growth through until spring; and if they do, they may start flowering half-

heartedly in July, although this is not always convenient. Cut the lot to the ground in March to give them a co-ordinated September start.

There are only four species, of which one is hardly grown at all, with one much less common and a little less hardy than the other two. This latter is *C. griffithii*, but my only experience of it has been to witness its demise in its first winter.

*Ceratostigma plumbaginoides* and *C. willmottianum* are gems; the former can most easily be considered as a smaller, more truly herbaceous version of the latter. Both have stiff, slender stems culminating in small heads of brilliant blue flowers in October just as the foliage turns increasingly fiery in colour; the blue flowers glint brightly among the red foliage. The former in

*Helianthus salicifolius* may be an enormous plant, but its emerging leaves and its bright flower heads are very attractive

particular has a tendency to creep at the root, but conversely if these wiry roots are pulled up and potted, they die.

Gardeners suspicious of begonias as sound autumn perennials will perhaps be a little more justifiably sceptical of *Impatiens tinctoria*. This is not just another busy lizzie, far from it. A 4–5ft herbaceous perennial with the single drawback of requiring a warm, sunny, sheltered wall to keep it growing long enough without frost for the flowers to come good. It grew, it may do still, in a border alongside a greenhouse in the botanic garden in Edinburgh, where the large, white flowers, each with a bold purplish maroon blotch on the lower petals, came as quite a shock. My garden is too cold for it; if yours is warmer try it.

Yet another confusing case, shrub or herbaceous perennial, is *Lespedeza thunbergii*, a plant seen more often, I think, in the United States than in Britain. You could think of it as a sort of herbaceous indigofera, with elegant arching growth and strings of small, purple-pink pea-flowers among the slightly silvery foliage from late September onwards. There is a white form and also one which produces pink, white and bicoloured flowers on the same plant. Well-drained soil, sun and a ferocious decapitation at ground level in spring (if the winter has not saved you the trouble) is all it requires.

Around the base of your lespedeza, liriopes can be grouped. Known in the States as lilyturf, *Liriope muscari*, muscari from the general shape of the flower spikes, which resembles those of grape hyacinths, is the one we all grow – I say all, I mean few; although seen filling beds in botanic gardens and National Trust gardens, in private gardens it is less common. But it's tough and reliable, unfailingly flowers in its slightly odd, pale violet shade in October, given sun and a reasonable soil, and its waterfall of narrow evergreen leaves is always valuable.

There is a white, 'Munroe White', which I find very slow, but it might thrive more contentedly in a sunnier, drier spot; I must move it. *Liriope spicata* is coarser and tougher and this too has a white form. The Americans and Japanese have more, including variegated ones.

Now, poke; yes, poke. This is a noun rather than an instruction, and is derived from the native American word *pak*, meaning blood and presumably referring to the juice from the fruits. It is known in the southern States as the basis of poke salad and was, as the saying goes, 'cooked in two waters' with a little fat pork and vinegar in the second. But it's worth growing simply so you can point to it and say to your visitor: 'Poke.' I can only say that the word refers to *Phytolacca americana*, an extraordinary plant used, among other things, for poisoning snails in fishponds without harming the fish! I understand you can also eat the young spring shoots, after two changes of water to remove the toxins, and though I've eaten stinging nettles, nasturtium tubers and the fleshy bits of yew berries in my time, this is a delicacy which I have yet to sample.

Poke is a stout plant, with unremarkable foliage, which gives little hint of its value until well into the summer when spikes of little white flowers appear. But it's hardly worth growing just for these. Later, in September and October, two things happen. First the berries start to appear, each spike like a cob of sweet corn packed tight with juicy blackcurrants. At the same time the stems develop tendencies towards the same shades and the leaves too... What a sight, a 5ft bush of purple and green. A hundred years ago Robinson reported a form with the leaves variegated in rose and white; it would surely catch the eye, but would we recoil or rejoice?

There is another species, *P. polyandra* (formerly *P. clavigera*), from China; its spikes of fruit are upright rather than drooping and its autumn leaves less colourful but if you have the space, grow the two. There is another, *P. acinosa*; I have not seen it.

Now that I list all the autumn salvias I wish to mention, and see the number of them, I realize how enormous my nascent autumn border is going to have to be. There's *Salvia azurea, S. guaranitica, S. leucantha, S. pitcheri, S. uliginosa* and *S. vanhouttei,* not to mention tender ones like *S. rutilans* and shrubby ones like *S. greggii* in its various forms. Oh dear, and we still have *Sedum* and *Tricyrtis* to come.

The very name of *Salvia azurea* is tempting, but it turns out to be little more than a smaller-flowered version of *S. pitcheri*. Both are over 3ft tall and will collapse on to the schizostylis you have thoughtfully planted in front unless discreetly staked. The mass of greyish foliage topped with dense heads of sharp blue flowers has that airiness so necessary among the more solid chrysanths and asters. Until recently *S. guaranitica* was rarely seen, yet now we have at least two listed and it keeps on turning up in the gardens of those who have the time to slip in a little brushwood to support its 5ft branching stems with their royal blue flowers. *S. leucantha* is different again, with the whole plant save the topsides of the leaves covered in silver wool. The flowers are white, but the whole effect is made by the pale dusky maroon calyces from which they emerge.

Of all these autumn salvias, *S. uliginosa* is perhaps the best known. The tall, widely branched stems with their clusters of sparkling blue, white-eyed flowers sway nonchalantly at the back of the border over the asters and the chrysanths which have, quite correctly, been set in front. Then *S. vanhouttei,* a plant I know only by repute but whose long dark red flowers out of red calyces sound valuable.

I could hardly complete this chapter without reference to the autumn sedums, so essential to both gardener and butterfly alike. Butterflies need their copious supplies of nectar before winter hibernation, we need the colour and the habit.

The slightly floppy *Sedum telephium* 'Atropurpureum' starts rather too early, in mid August, with its smoky pink flowers set against purple foliage on glossy stems, and the many forms of *Sedum spectabile* are the mainstays of the group.

In fact one of the best known of these autumn sedums, 'Autumn Joy', is a hybrid between *S. telephium* and *S. spectabile*. Its foliage is unremarkable, but the rich pink flowers in oddly bumpy heads attract plenty of bees but often no butterflies at all. At 2ft in height 'Autumn Joy', often now listed under its original name of 'Herbstfreude', is taller than the forms of *S. spectabile*, most of which manage only 15–18in.

'Brilliant' is the best of all these for butterflies; a clump in the Dell garden at Bressingham had collected more butterflies per square foot than I've ever seen in my life when I checked them all one September. The flowers are pale pink but the foliage is very washed out. The leaves of 'Iceberg' are as unappealing but the white heads, while smaller than those of most varieties, hold their colour well before fading to cream. Occasionally a pink-flowered shoot appears.

The best of the rest are 'Meteor' and 'Indian Chief'. All these sedums are best split before the clumps get too congested, otherwise the shoots become so crowded that a downpour bends them over and they do not always recover. 'Meteor' seems better able to withstand the wind and the rain; it supports itself more stoutly, but the foliage is the least good of any – sometimes almost cream. The pale flowers open from silvery white buds and eventually darken; in second place for bees and butterflies. 'Indian Chief' is a

179

little later than 'Meteor'; its foliage is far superior, in the usual glaucous sedum colouring, and the unusually silvery buds open to brilliant pink flowers with masses of bees but fewer butterflies.

Finally come those grown as much for their foliage as their flowers. In truth, the foliage is more important, for none of the dark-leaved varieties have flowers to match 'Brilliant', 'Meteor' and the others.

'Arthur Branch', a form of *S. telephium*, has the richest purple leaves, on red stems too, with pale purple flowers and I think I like this the best. 'Munstead Red' has better, darker flowers but the foliage is no more than green tinted with purple. 'Vera Jameson' has watery, livery-purple leaves and brownish flowers but this is in October, it's better earlier. 'Roseovariegatum' I have not seen and I will not be rushing to inspect it: 'new shoots in spring become strongly suffused in bright pink and almost stop growing'. Fortunately they then turn green and growth accelerates. 'Gooseberry Fool' was picked out by Graham Thomas and is unique in its purple mark on each green leaf; that sounds more appealing.

And now, at the end of this far too lengthy chapter, a few tricyrtis. It cannot surely be their common name of toad lily which puts gardeners off these intriguing plants; I would plant them in broader drifts if they were more like toads. There must be some evil rumour abroad that they are far more difficult to grow than they actually are; roscoeas suffer the same reputation. Partial shade and soil which though not soggy does not dry out are all they require, and there can hardly be a garden in the country in which such conditions cannot be provided. And where, I might say, a great number of other delightful plants can be suited.

For tricyrtis can be planted among any number of spring plants such as snowdrops and primroses, will in no way hamper their growth or impede their flowering, and in autumn, when these earlier flowers have vanished or are irritatingly ragged, will overtop them all and burst into bloom.

Not that they are especially flamboyant; hardly. But their delicate spotting places them among the most individually beautiful of autumn flowers. They have been described as 'weird' and it has been said that they 'take some getting used to'; but for me, one look was enough. Some may be 3ft tall, but never be tempted to place them at the back of the bed; they must be at the front so that their spotting may be inspected closely. The flowers of the creeping *T. formosana* are almost white with dense purple spotting and gathered in heads at the tops of the stems. Those of the hairy-leaved *T. hirta* line the top half of the plant, bursting out of the leaf axils, and are delicately spotted; the rootstock is also more compact. *Tricyrtis latifolia* is yellow spotted with brown and the leaves are broad.

'White Towers' is a pure white hybrid, reckoned to have *T. hirta* and *T. affinis* as parents, and lovely for cutting, as they all are, with arching stems from which the flowers turn up prettily. Occasionally seedlings may appear in lilac or yellow without spots, and where you grow more than one species you may find hybrids.

So ends this autumn choice. I find that having reached this point I could almost turn over the whole garden to autumn flowers – were it not for feeling just the same about winter and spring. It is heartening, though, that this neglected season has so many wonderful flowers waiting to rescue it from obscurity.

# AFTERWORD

'I have written a good lot and as I should think nearly sufficient,' as the poet John Clare put it; but perhaps not *quite* sufficient. It seems customary for authors first to write their book, then at the end of that process to write the introduction to the book in a whirl of relief – usually on the very day when the publisher demands the final manuscript. In this case I wrote the introduction when I was half-way through, but now feel that after that last rather hectic spin, the tumble of all those autumn perennials, a little peace, a little restfulness is required; hence this afterword.

In the garden, an atmosphere of quiet repose is created by seclusion, symmetry and green foliage. The one thing provided by neither the traditional herbaceous border nor the modern mixed border is this peace – there is simply too much colour, too many flowers demanding a reaction. Perhaps this explains why woodland glades, after the colourful pattern of the spring mosaic has faded, are especially restful. The dark leafy canopy of deciduous trees, the interlacing leaves of hostas, ferns and iris or sedge, predominantly green, and the absence of sharp colours to jag at our tranquil spirit allow us to drink in the stillness – and relax.

Emerging from the cool of the summer glade the colours of the beds and borders seem sharper, more real. This, of course, is the purpose of the lawn, a pool of unruffled calm against which to view the brilliant borders. Every garden needs tranquillity, and while this whole book is about plants and their diversity it is perhaps wise to conclude with the suggestion that a glade of visual and atmospheric stillness, with no special plants at all, will not only help restore the gardener's spirit but enhance our appreciation of all those plants which crowd the rest of the garden.

*Opposite*

'Purleigh White' is one of a small number of neat, hardy, pompom chrysanthemums descended from the pink 'Mei-Kyo', cuttings of which first arrived in Britain from Japan in a matchbox.

*Overleaf (left)*

ABOVE At Washfield Nursery the recently rediscovered *Aster novi-belgii* 'Calliope', with its distinctive dark stems, falls forward into the developing seed-heads of the dwarf *Miscanthus sinensis* 'Kleine Fontäne'.

BELOW LEFT *Liriope muscari* is sometimes dismissed as nothing more than a ground cover plant and ignored but, as here at Wisley, its flowering capacity reveals it as a valuable autumn perennial.

BELOW RIGHT *Anemone Blanda* 'White Splendour' has large, pure white flowers, with pink tints to the backs of the petals which are especially obvious in the buds. It makes vigorous cover under deciduous shrubs.

*Overleaf (right)*

ABOVE The irresistible markings of *Tricyrtis formosana* 'Stolonifera' are best seen at the front of the border, even though the plant would normally be considered too tall for such a situation.

BELOW It may be a surprise to find that there are frost-hardy begonias, but *B. grandis* var. *evansiana*, seen here at Bressingham's Dell Garden, is a valuable autumn perennial and as hardy as many other perennials in chilly gardens.

# HARDLY A BIBLIOGRAPHY

The more revolutionary of my readers, those who have read this book by starting at the beginning and continuing until this final chapter, will have noticed that I too have been reading. When Annie Lee, my indispensable copy-editor who ensures that my double negatives mean what I need them to mean and not what everyone thinks they can't possibly mean, when she suggested a final, *final* chapter listing all the books that seem to have been on my mind when writing this book – well, I usually take her advice so here goes.

CLAUDE A. BARR, *Jewels of the Plains* (University of Minnesota Press, 1983). Now out of print, this is an evocative examination of the plants of the Great Plains by a rancher from South Dakota turned nurseryman. His first catalogue, issued in 1935, was entitled 'Beautiful Native Plants from the High Plains, Badlands and Black Hills'!

RICHARD BIRD, *Garden Pinks* (Batsford, 1994). A most valuable modern account by one of our most prolific authors on hardy perennials. Look out too for *The Propagation of Hardy Perennials* (Batsford, 1993) and *The Cultivation of Hardy Perennials* (Batsford, 1994).

*Opposite*
One of the classic autumn perennials is *Aster* x *frikartii* 'Mönch', seen here at Old Court in Worcestershire. Distinguished by its almost flat ray petals, it also has an unusually long season, often flowering from early summer until late autumn.

FRANK BISHOP, *The Delphinium* (Collins, 1949). Valuable as a snapshot of the post-war delphinium world but low on inspiration.

E. A. BOWLES, *My Garden in Spring, My Garden in Summer, My Garden in Autumn and Winter* (T. C. and E. C. Jack, 1914 and 1915). Reprinted by David & Charles in 1972 but now again out of print, these three books contain some of the keenest observation and most entertaining writing about plants that you will find anywhere.

COLETTE, *Flowers and Fruit* (Farrar, Straus, Giroux, 1986). This selection of short plant portraits is made from a large body of Colette's non-fiction, of which there is actually more than there is of the better-known fiction. Some were written and published in Nazi-occupied Paris to help her afford black-market cheese and chicken, but their romantic perception is undimmed.

CANON HENRY ELLACOMBE, *In a Gloucestershire Garden* (Edward Arnold, 1895). Republished by Century in 1982, this month-by-month account is full of thoughtful observation lightened with occasional eccentricities, like the inclusion of the poem by C. Mackay on the Christmas rose.

REGINALD FARRER, *My Rock-Garden*, (Edward Arnold, 1907) and *The English Rock-Garden* (Nelson, 1919). These two books include plenty of plants now grown as perennials, and it's only occasionally that the extraordinarily hyperbolic writing gets the better of the great man's botanical instincts.

MARGERY FISH. Of her eight books everyone has

their own favourites; mine are *An All Year Garden* (David & Charles, 1958), *Gardening in the Shade* (Collingridge, 1964), *Ground Cover Plants* (Collingridge, 1964) and *A Flower for Every Day* (Studio Vista, 1965). Most of Mrs Fish's books were reissued more recently in paperback by Faber, but are out of print again at the time of writing. All are firmly based on her experience at East Lambrook, a garden now restored using her writings as a guide.

FLORA EUROPAEA (Cambridge University Press). This five-volume work is in a constant state of revision and is an invaluable guide to all the plants of Europe for the botanically-minded.

ROY GENDERS. This extraordinarily prolific author wrote 100 books, including four on primulas. These were *Primroses and Polyanthus*, with H. C. Taylor (Faber, 1954), *Auriculas* (John Gifford, 1958), *Primroses* (John Gifford, 1959) and *The Polyanthus* (Faber, 1963). All are long out of print but they contain practical and historical information not easily found elsewhere; unfortunately they are not always 100 per cent reliable.

DIANA GRENFELL, *Hosta, the Flowering Foliage Plant* (Batsford, 1990). This is all most gardeners will require on hostas – its descriptive list of varieties, while by no means comprehensive, is most valuable. There are useful lists of varieties for different purposes.

GEOFFREY GRIGSON. His two books on the names of British wild plants and their origins, *The Englishman's Flora* (Phoenix House, 1958) and *A Dictionary of English Plant Names* (Allen Lane, 1974), are among the most fascinating of all books on the British flora.

ROGER GROUNDS, *Ornamental Grasses* (Helm, 1989). This most comprehensive of books may

be a little difficult for newcomers to grasses to digest, but all the information is there.

ALICE HOFFMAN. The heading for the third main section has been pinched from the novel of the same name by Alice Hoffman, one of America's finest novelists.

GERTRUDE JEKYLL. Her books, originally published by Country Life and more recently reissued by the Antique Collectors' Club, are not all as inspired as some of her fans make out, or as perfectly written. However, all are worth reading and the recent version of *Colour Schemes for the Flower Garden*, published by Frances Lincoln (1988) with newly drawn plans, is the best one with which to start.

LEO JELLITO AND WILHELM SCHACHT. Their two-volume *Hardy Herbaceous Perennials* (Batsford, 1990) is usually my first reserve when looking up perennials, after Graham Thomas (see below). The descriptions are much more detailed, but fewer plants are included and the nomenclature is out of date.

ALLEN LACY, *The Garden in Autumn* (Atlantic Monthly Press, 1990). Not long in print and never available in Britain, this is a most elegantly written and inspiring book. He too has noticed that the poets got it wrong about autumn. *Home Ground* and his other anthologies are also worth hunting out.

PETER LEWIS AND MARGARET LYNCH, *Campanulas* (Helm, 1989). This monograph, the only recent one on the subject, is one of the best of recent years, particularly in the way it presents a great deal of detailed information in a very readable way.

CHRISTOPHER LLOYD. Most of his writing is worth reading. *Hardy Perennials* (Studio Vista, 1967) is long out of print, as are some of his other early books, but of those still available I would pick the expanded versions of *The Well-Tempered Garden* (Viking, 1985) and *Foliage Plants* (Viking, 1985), together with *In My Garden* (Bloomsbury, 1993).

TONY LORD AND CHRIS PHILLIP, *The Plant Finder* (Moorland/RHS). This annual publication, listing over 60,000 plants and where to buy them, has also set the standard of nomenclature which everyone is inclined to follow.

ANN LOVEJOY, *The Border in Bloom* (Sasquatch Books, 1990). To be honest, I don't think I've included anything relating to this most enjoyable book; but I should have done. Ann Lovejoy doesn't care, she writes what she thinks – which would be unbearable if she never had an original thought. Fortunately, her writing derives directly from her own wit and her thoughtful observation and experience – so we're safe.

HENRY MITCHELL. One of America's most thoughtful and provocative writers, now no longer with us, *The Essential Earthman* and *One Man's Garden* are published by Houghton Mifflin in the States and Cassell in the UK. Hunt them out.

NCCPG, *The Pink Sheet* (1991). This list of 1,500 rare and endangered garden plants was prepared by the Cambridge Group of the National Council for the Conservation of Plants and Gardens and may cause the hearts of plantaholics to miss more beats than just the one.

JOHN PARKINSON, *Paradisi in Sole: Paradisus Terrestris*. Published in 1629, this was the first of the great British books on flowers, with an extraordinary variety of plants described and illustrated with woodcuts. Republished in paperback in 1991 as *A Garden of Pleasant Flowers* at a tempting price by Dover Publications.

GEORGE A. PHILLIPS, *Delphiniums* (Thornton Butterworth, 1933). Long out of print, this is good on history and period varieties and is the most inspiring of all delphinium books; written with real enthusiasm.

C. T. PRIME, *Lords and Ladies* (Collins, 1960). One of those unique Special Volumes in the New Naturalist series. I know I've treated it with a certain sarcasm but it really is a classic of painstaking research. Republished by the author's widow.

GRAHAM RICE AND ELIZABETH STRANGMAN, *The Gardener's Guide to Growing Hellebores* (David & Charles, 1993). Covers the species and cultivars in accessible detail, with additional focus on the details of cultivation.

WILLIAM ROBINSON, *The English Flower Garden* (John Murray, 1883). An extraordinary book, only part of it actually written by the eccentric Robinson, and changing from one edition to another over many years. Still a good reference and a great read.

ROYAL HORTICULTURAL SOCIETY, *New Dictionary of Gardening* (Macmillan/RHS, 1992). An extraordinary feat of publishing, but showing the result of rushing the process.

VITA SACKVILLE-WEST, *The Garden Book* (Michael Joseph, 1968). Her pieces originally appeared in the *Observer* in the 1950s then in collections entitled *In Your Garden, In Your Garden Again, More for Your Garden* and finally, you've guessed it, *Even More for Your Garden*. Then this further selection from these selections appeared in various editions.

W. GEORG SCHMID, *The Genus Hosta* (Batsford, 1991). Notwithstanding my facetiousness, this is an extraordinarily comprehensive book for hosta fanatics.

SACHEVERELL SITWELL, *Old Fashioned Flowers* (Country Life, 1939). Fascinating account of the florists' flowers, including carnations, pinks, primroses and dahlias with illustrations by John Farleigh.

B. H. B. SYMONS-JEUNE, *Phlox* (Collins, 1954). The only book on border phlox and valuable as far as it goes, but with rather too much emphasis on his own varieties.

GRAHAM STUART THOMAS, *Perennial Garden Plants* (Dent, 1976). A perennial classic revised in 1990 and issued in paperback, this intentionally subjective book looks a little dated in view of the vast flow of new introductions of the late 1980s and 1990s; but it is still the first place to look.

PIERS TREHANE, *Index Hortensis: Volume I: Perennials* (Quarterjack, 1989). A detailed guide to the correct names of the hardy perennials in cultivation with references to help find more information.

ROGER TURNER, *Hardy Euphorbias* (Batsford, 1995). The only book on these popular plants; extremely thorough and inspiring.

HELEN VAN PELT WILSON, *The New Perennials Preferred* (Collier, 1992). First appearing in 1945 and extended in 1961, this long series of short essays, season by season, has lasted unusually well.

LOUISE BEEBE WILDER, *Colour in My Garden* (Doubleday, 1918). A writer almost unknown in the UK but rightly respected in the USA. This book is witty, inspiring and a joy to read. She specialized in unlikely titles: her other books include *What Happens in My Garden*, an appealingly downbeat title, the unlikely *Adventures with Hardy Bulbs* and what sounds like a Famous Five romp, *Adventures in a Suburban Garden*.

PETER YEO, *Hardy Geraniums* (Helm, 1985) This splendid book has the unlikely distinction of selling well through book clubs, which must make it unique among monographs by professional taxonomists. An outstanding treatment of a popular plant.

# INDEX

Page numbers in *italic* refer to the illustrations